McGraw

# 500 College Physics Questions

**Also in McGraw-Hill's 500 Questions Series**

*McGraw-Hill's 500 American Government Questions: Ace Your College Exams*
*McGraw-Hill's 500 College Algebra and Trigonometry Questions: Ace Your College Exams*
*McGraw-Hill's 500 College Biology Questions: Ace Your College Exams*
*McGraw-Hill's 500 College Calculus Questions: Ace Your College Exams*
*McGraw-Hill's 500 College Chemistry Questions: Ace Your College Exams*
*McGraw-Hill's 500 College Physics Questions: Ace Your College Exams*
*McGraw-Hill's 500 Differential Equations Questions: Ace Your College Exams*
*McGraw-Hill's 500 European History Questions: Ace Your College Exams*
*McGraw-Hill's 500 French Questions: Ace Your College Exams*
*McGraw-Hill's 500 Linear Algebra Questions: Ace Your College Exams*
*McGraw-Hill's 500 Macroeconomics Questions: Ace Your College Exams*
*McGraw-Hill's 500 Microeconomics Questions: Ace Your College Exams*
*McGraw-Hill's 500 Organic Chemistry Questions: Ace Your College Exams*
*McGraw-Hill's 500 Philosophy Questions: Ace Your College Exams*
*McGraw-Hill's 500 Physical Chemistry Questions: Ace Your College Exams*
*McGraw-Hill's 500 Precalculus Questions: Ace Your College Exams*
*McGraw-Hill's 500 Psychology Questions: Ace Your College Exams*
*McGraw-Hill's 500 Spanish Questions: Ace Your College Exams*
*McGraw-Hill's 500 U.S. History Questions, Volume 1: Ace Your College Exams*
*McGraw-Hill's 500 U.S. History Questions, Volume 2: Ace Your College Exams*
*McGraw-Hill's 500 World History Questions, Volume 1: Ace Your College Exams*
*McGraw-Hill's 500 World History Questions, Volume 2: Ace Your College Exams*
*McGraw-Hill's 500 MCAT Biology Questions to Know by Test Day*
*McGraw-Hill's 500 MCAT General Chemistry Questions to Know by Test Day*
*McGraw-Hill's 500 MCAT Physics Questions to Know by Test Day*

McGraw-Hill's

# 500 College Physics Questions

*Ace Your College Exams*

Alvin Halpern, PhD

New York Chicago San Francisco Lisbon London Madrid Mexico City
Milan New Delhi San Juan Seoul Singapore Sydney Toronto

The McGraw-Hill Companies

**Dr. Alvin Halpern's** teaching experience includes chairing the physics department at Brooklyn College, the City University of New York.



1 2 3 4 5 6 7 8 9 10 11 12 13 14 15 16 17 QFR/QFR 1 9 8 7 6 5 4 3 2

ISBN 978-0-07-178982-0
MHID 0-07-178982-0



e-ISBN 978-0-07-178983-7
e-MHID 0-07-178983-9

Library of Congress Control Number 2011944587



This book is printed on acid-free paper.

# CONTENTS

# INTRODUCTION

You've taken a big step toward success in physics by purchasing *McGraw-Hill's 500 College Physics Questions*. We are here to help you take the next step and score high on your first-year exams!

This book gives you 500 exam-style questions that cover all the most essential course material. Each question is clearly explained in the answer key. The questions will give you valuable independent practice to supplement your regular textbook and the ground you have already covered in your class.

This book and the others in the series were written by experienced teachers who know the subject inside and out and can indentify crucial information as well as the kinds of questions that are most likely to appear on exams.

You might be the kind of student who needs to study extra before the exam for a final review. Or you might be the kind of student who puts off preparing until the last minute before the test. No matter what your preparation style, you will benefit from reviewing these 500 questions, which closely parallel the content and degree of difficulty of the questions on actual exams. These questions and the explanations in the answer key are the ideal last-minute study tool.

If you practice with all the questions and answers in this book, we are certain you will build the skills and confidence needed to excel on your exams.

*—Editors of McGraw-Hill Education*

McGraw-Hill's

# 500 College Physics Questions

CHAPTER 1

# Equilibrium of Concurrent Forces

## Planar Vectors, Scientific Notation, and Units

**1.** What is a scalar quantity?

**2.** What is a vector quantity?

**3.** Describe the graphical addition of vectors.

**4.** What is a *component* of a vector?

## Ropes, Knots, and Frictionless Pulleys

**5.** The object in Figure 1.1 weighs 50 N and is supported by a cord. Find the tension in the cord.

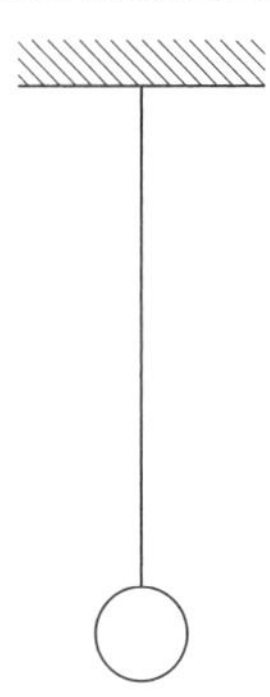

**Figure 1.1**

**6.** As shown in Figure 1.2, the tension in the horizontal cord is 30 N. Find the weight ($w$) of the object.

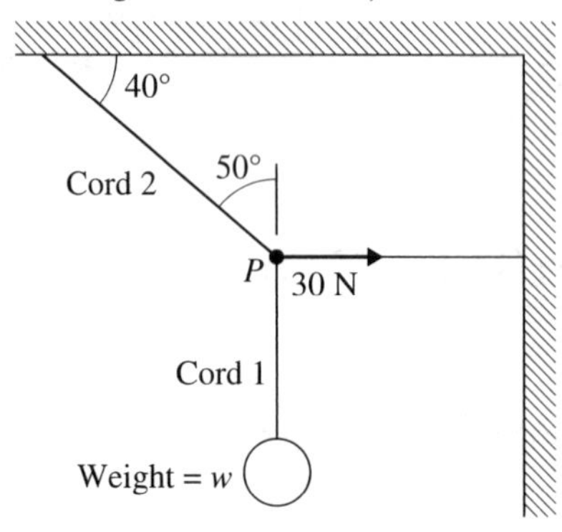


**Figure 1.2**

**7.** If the identical ropes $A$ and $B$ can each support tensions no greater than 200 N, what is the maximum value that $W$ can have? What is the tension in the other rope when $W$ has this maximum value?

**8.** A rope extends between two poles. A 90-N boy hangs from it. Find the tensions in the two parts of the rope.

**9.** If $w = 40$ N in the equilibrium situation shown in Figure 1.3, find $T_1$ and $T_2$.

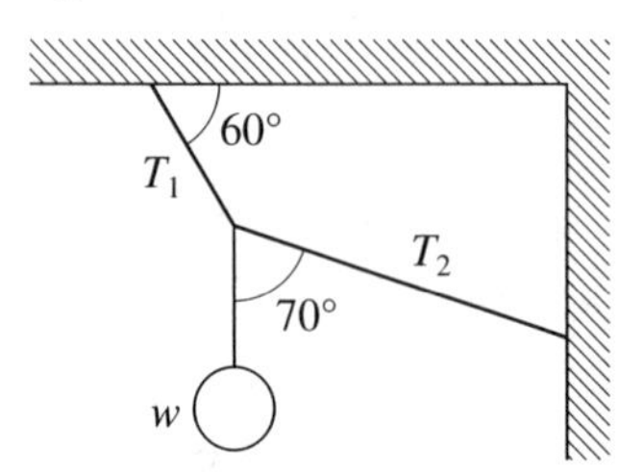


**Figure. 1.3**

## Friction and Inclined Planes

**10.** A 200-N wagon is to be pulled up a 30° incline at constant speed. How large a force parallel to the incline is needed if friction effects are negligible?

**11.** A box weighing 100 N is at rest on a horizontal floor. The coefficient of static friction between the box and the floor is 0.4. What is the smallest force $F$ exerted eastward and upward at an angle of 30° with the horizontal that can start the box in motion?

**12.** A 50-N box is slid straight across the floor at constant speed by a force of 25 N, as shown in Figure 1.4. How large a friction force ($f$) impedes the motion of the box? How large is the normal force?

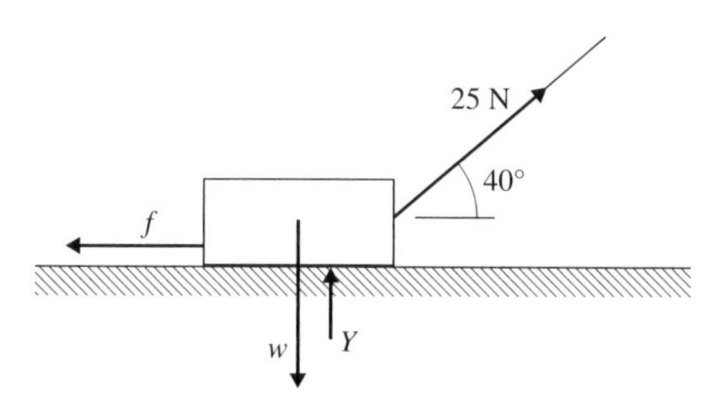


**Figure 1.4**

**13.** Each of the objects in Figure 1.5 is in equilibrium. Find the normal force, $Y$, in each case.

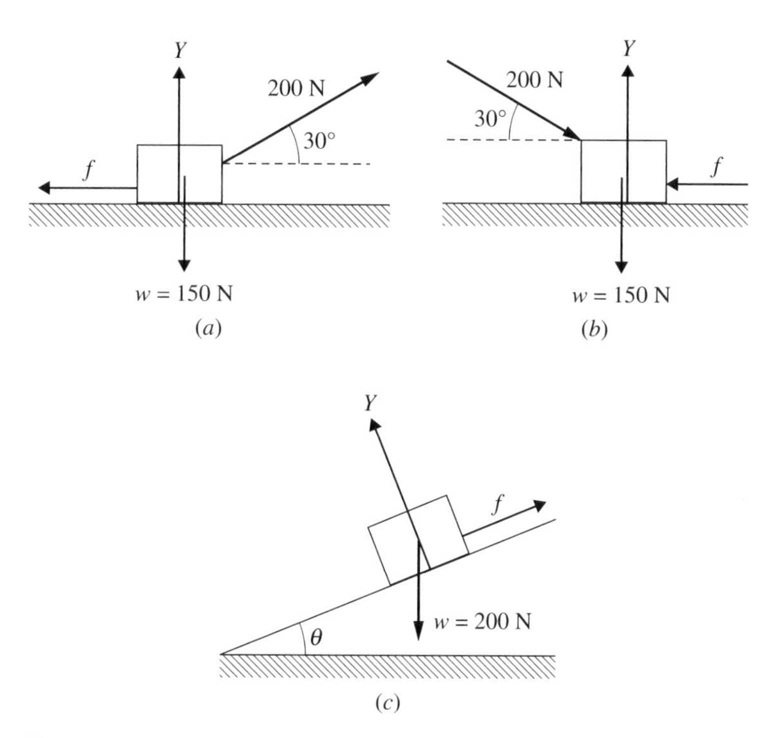


**Figure 1.5**

**14.** Assume that $W = 60$ N, $\theta = 43°$, and $\mu_k = 0.3$ in Figure 1.6. What push will move the block with constant speed?

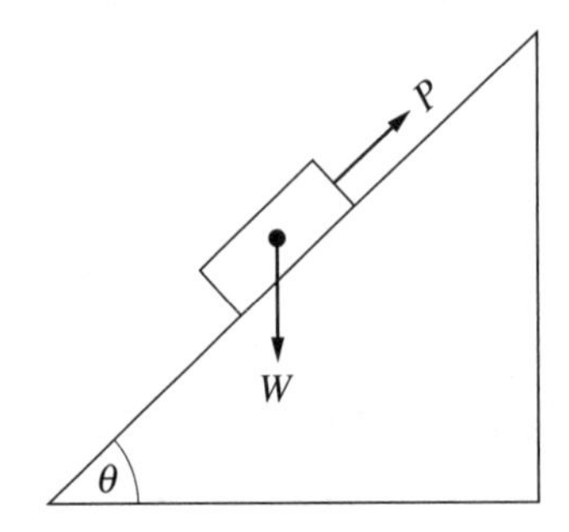


**Figure 1.6**

(A) up the plane
(B) down the plane

**15.** A block slides with constant speed under the action of the force shown (see Figure 1.7).

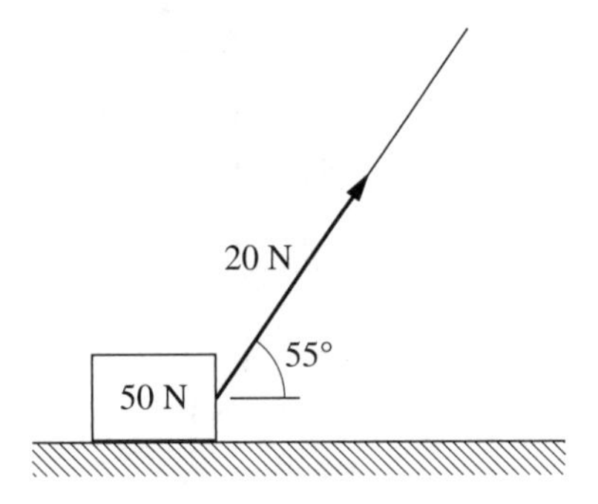


**Figure 1.7**

(A) How large is the retarding friction force?
(B) What is the coefficient of kinetic friction between the block and the floor?

**16.** A block slides at constant speed down the incline (see Figure 1.8).

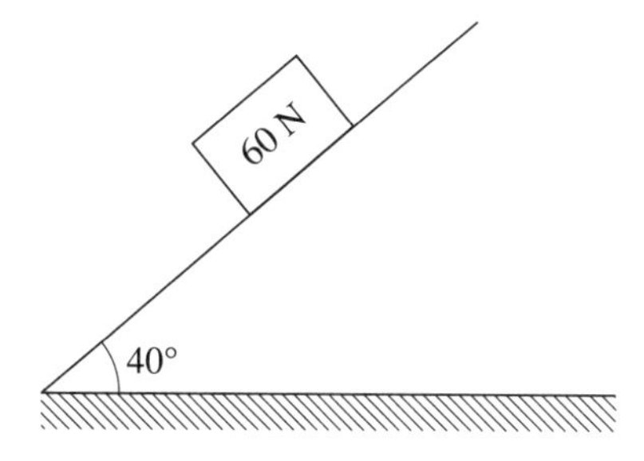


**Figure 1.8**

(A) How large is the friction force that opposes its motion?
(B) What is the coefficient of sliding (kinetic) friction between the block and plane?

CHAPTER 2

# Kinematics in One Dimension

## Dimensions and Units; Constant-Acceleration Problems

**17.** A rocket-propelled car starts from rest at $x = 0$ and speeds up with constant acceleration $a = 5$ m/s$^2$ for 8 s until the fuel is exhausted. It then continues with constant velocity. What distance does the car cover in 12 s?

**18.** The particle shown in Figure 2.1 moves along $x$ with a constant acceleration of –4 m/s$^2$. As it passes the origin, moving in the + direction of $x$, its velocity is 20 m/s. In this problem, time $t$ is measured from the moment the particle is first at the origin.

(A) At what distance $x'$ and time $t'$ does $v = 0°$?
(B) At what time is the particle at $x = +15$ m, and what is its velocity at this point?
(C) What is the velocity of the particle at $x = +25$ m? at $x = -25$ m?
(D) Why is it not possible to find the velocity of the particle at $x = +55$ m?

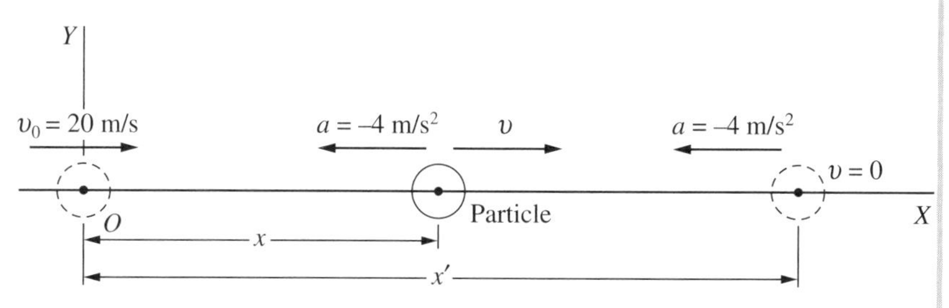


**Figure 2.1**

**19.** A body falls freely from rest. Find

(A) its acceleration
(B) the distance it falls in 3 s
(C) its speed after falling 70 m
(D) the time required to reach a speed of 25 m/s
(E) the time taken to fall 300 m

**20.** A ball is thrown vertically downward from the edge of a high cliff with an initial velocity of 25 ft/s.

(A) How fast is it moving after 1.5 s?
(B) How far has it moved after 1.5 s?

**21.** A stone is thrown downward with initial speed 8 m/s from a height of 25 m. Find

(A) the time it takes to reach the ground
(B) the speed with which it strikes

**22.** A ball thrown vertically upward returns to its starting point in 4 s. Find its initial speed.

**23.** A ballast bag is dropped from a balloon that is 300 m above the ground and rising at 13 m/s. For the bag, find

(A) the maximum height reached
(B) its position and velocity 5 s after being released
(C) the time before it hits the ground

**24.** The acceleration due to gravity on the moon is 1.67 $m/s^2$. If a person can throw a stone 12.0 m straight upward on the earth, how high should the person be able to throw a stone on the moon? Assume that the throwing speeds are the same in the two cases.

**25.** A proton in a uniform electric field moves along a straight line with constant acceleration. Starting from rest, it attains a velocity of 1000 km/s in a distance of 1 cm.

(A) What is its acceleration?
(B) What time is required to reach the given velocity?

**26.** A bottle dropped from a balloon reaches the ground in 20 s. Determine the height of the balloon if

(A) it was at rest in the air
(B) it was ascending with a speed of 50 m/s when the bottle was dropped

## Graphical and Other Problems

**27.** The graph of an object's motion (along a line) is shown in Figure 2.2.
 (A) Find the instantaneous velocity of the object at points $A$ and $B$.
 (B) What is the object's average velocity between points $A$ and $B$?
 (C) What is the object's average acceleration between points $A$ and $B$?

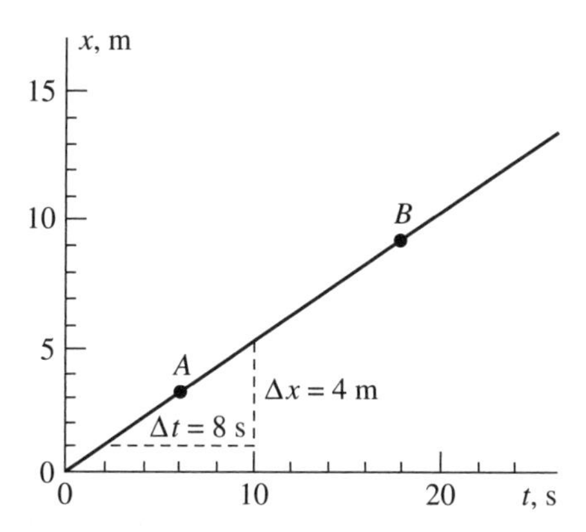


**Figure 2.2**

**28.** Refer to Figure 2.3. Find the instantaneous velocity at point $F$ for the object whose motion the curve represents.

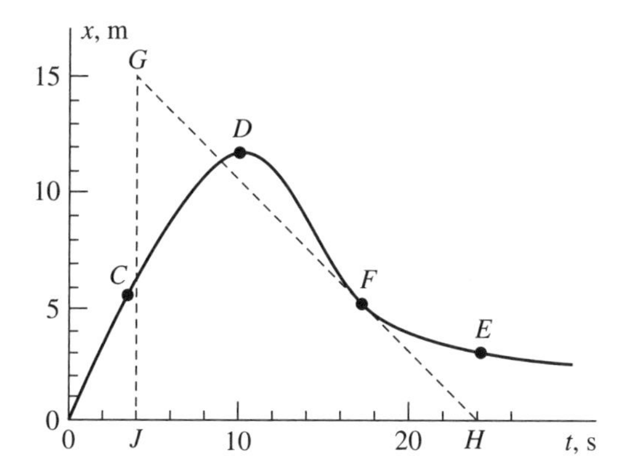


**Figure 2.3**

**29.** A girl walks along an east-west street, and a graph of her displacement from home is shown in Figure 2.4. Find her average velocity for the whole time interval shown as well as her instantaneous velocity at points *A*, *B*, and *C*.

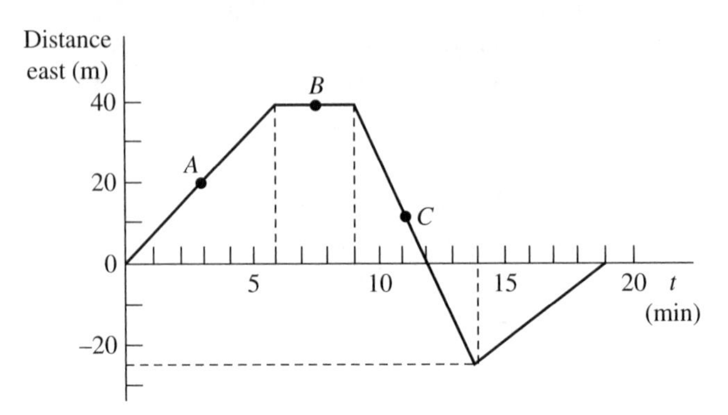


**Figure 2.4**

**30.** Referring to Figure 2.4, find

(A) the average velocity for the time interval $t = 7$ min to $t = 14$ min
(B) the instantaneous velocity at $t = 13.5$ min
(C) the instantaneous velocity at $t = 15$ min

**31.** The graph of a particle's motion along the $x$ axis is given in Figure 2.5.

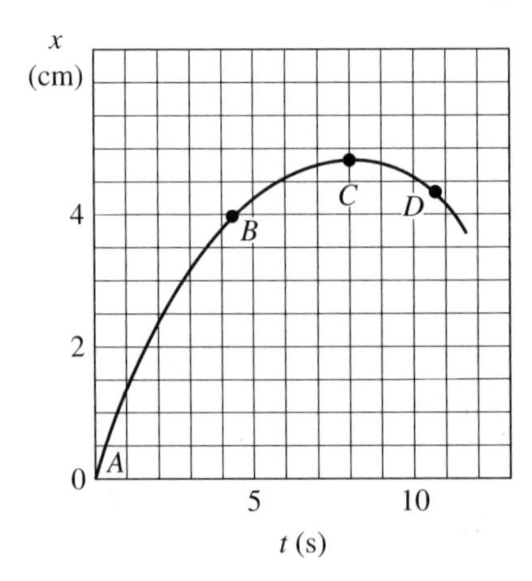


**Figure 2.5**

Estimate the

(A) average velocity for the interval from *A* to *C*
(B) instantaneous velocity at *D*
(C) instantaneous velocity at *A*

**32.** Figure 2.6 shows the velocity of a particle as it moves along the $x$ axis.

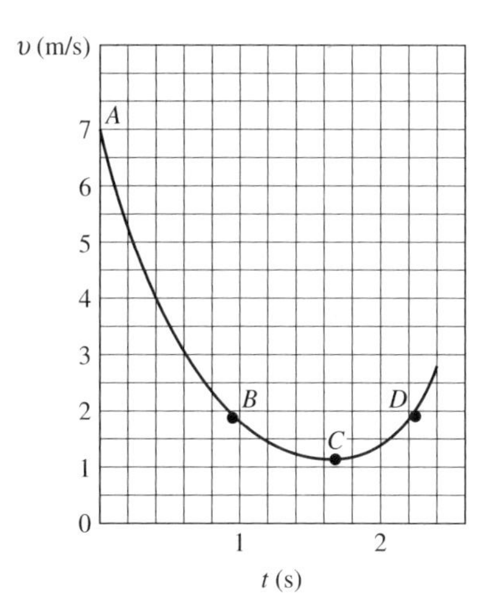


**Figure 2.6**

Find its acceleration at

(A) $A$
(B) $C$

**33.** A ball is thrown vertically upward with a velocity of 20 m/s from the top of a tower having a height of 50 m, as in Figure 2.7. On its return, it misses the tower and finally strikes the ground,

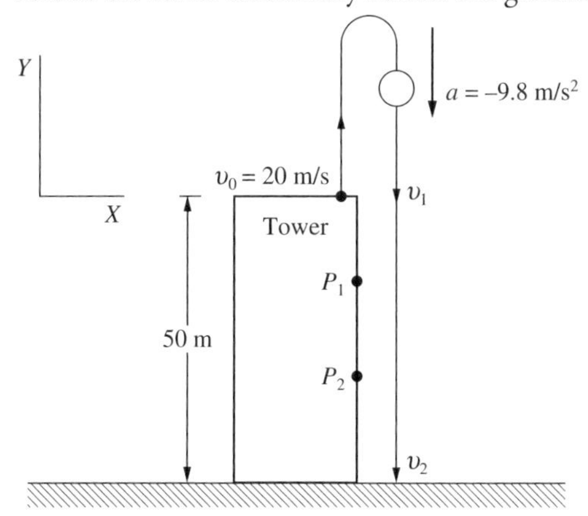


**Figure 2.7**

(A) what time $t_1$ elapses from the instant the ball was thrown until it passes the edge of the tower? What velocity $v_1$ does it have at this time?
(B) what total time $t_2$ is required for the ball to reach the ground? With what velocity $v_2$ does it strike the ground?

**34.** Refer to Question 33 and Figure 2.7

(A) What is the maximum height above ground reached by the ball?
(B) Points $P_1$ and $P_2$ are 15 and 30 m, respectively, below the top of the tower. What time interval is required for the ball to travel from $P_1$ to $P_2$?
(C) It is desired that after passing the edge, the ball will reach the ground in 3 s. With what velocity must it be thrown upward from the roof?

**35.** A man runs at a speed of 4.0 m/s to overtake a standing bus. When he is 6.0 m behind the door (at $t = 0$), the bus moves forward and continues with a constant acceleration of 1.2 m/s$^2$.

(A) How long does it take for the man to gain the door?
(B) If in the beginning he is 10.0 m from the door, will he (running at the same speed) ever catch up?

**36.** A ball is dropped from the top of a building. The ball takes 0.5 s to fall past the 3-m length of a window some distance from the top of the building.

(A) How fast was the ball going as it passed the top of the window?
(B) How far is the top of the window from the point at which the ball was dropped?

**37.** A truck is moving forward at a constant speed of 21 m/s. The driver sees a stationary car directly ahead at a distance of 110 m. After a "reaction time" of $\Delta t$, he applies the brakes, which gives the truck a backward acceleration of 3 m/s$^2$.

(A) What is the maximum allowable $\Delta t$ to avoid a collision, and what distance will the truck have moved before he applies the brakes?
(B) Assuming a reaction time of 1.4 s, how far behind the car will the truck stop, and in how many seconds from the time the driver first saw the car?

**38.** Just as a car starts to accelerate from rest with acceleration 1.4 m/s$^2$, a bus moving with constant speed of 12 m/s passes it in a parallel lane.

(A) How long before the car overtakes the bus?
(B) How fast will the car then be going?
(C) How far will the car then have gone?

**39.** A monkey in a perch 20 m high in a tree drops a coconut directly above your head as you run with speed 1.5 m/s beneath the tree.

(A) How far behind you does the coconut hit the ground.
(B) If the monkey had really wanted to hit your toes, how much earlier should the coconut have been dropped?

**40.** Two balls are dropped to the ground from different heights. One ball is dropped 2 s after the other, but they both strike the ground at the same time, 5 s after the first is dropped.

(A) What is the difference in the heights at which they were dropped?
(B) From what height was the first dropped?

**41.** Two boys start running straight toward each other from two points that are 100 m apart. One runs with a speed of 5 m/s, while the other moves at 7 m/s. How close are they to the slower one's starting point when they reach each other?

CHAPTER 3

# Newton's Laws of Motion

## Force, Mass, and Acceleration

**42.** Fill in the blanks.

(A) The mass of a 300-g object is __________.
(B) Its weight on earth is__________.
(C) An object that weighs 20 N on earth has a mass on the moon equal to __________.
(D) The mass of an object that weighs 5 N on earth is __________.

**43.** A 900-kg car is going 20 m/s along a level road. How large a constant retarding force is required to stop it in a distance of 30 m?

**44.** A 20-kg crate hangs at the end of a long vertical rope. Find its acceleration when the tension in the rope is

(A) 250 N
(B) 150 N
(C) zero
(D) 196 N

**45.** A 40-kg trunk sliding across a floor slows down from 5.0 to 2.0 m/s in 6.0 s. Assuming that the force acting on the trunk is constant, find its magnitude and its direction relative to the velocity vector of the trunk.

**46.** A resultant force of 20 N gives a body of mass $m$ an acceleration of 8.0 $m/s^2$, and a body of mass $m'$ an acceleration of 24 $m/s^2$. What acceleration will this force cause the two masses to acquire if they are fastened together?

**47.** A boy having a mass of 75 kg holds in his hands a bag of flour weighing 40 N. What is the normal force on the boy's feet?

**48.** Two drivers, one owning a large Cadillac and the other owning a small Volkswagen, make a bet. The VW owner bets that his car can pull as hard as the Cadillac. They chain the two rear bumpers together in a large empty parking lot. Each driver gets into his car and applies full power. The Cadillac pulls the VW backward all over the lot. The driver of the VW later claims that his car was pulling on the chain as hard as the Cadillac all the time. What does Newton's third law say in this case? Assume that the chain has negligible mass.

**49.** An elevator starts from rest with a constant upward acceleration. It moves 2.0 m in the first 0.60 s. A passenger in the elevator is holding a 3-kg package by a vertical string. What is the tension in the string during the accelerating process?

**50.** A boy who normally weighs 300 N on a bathroom scale crouches on the scale and suddenly jumps upward. His companion notices that the scale reading momentarily jumps up to 400 N as the boy springs upward. Estimate the boy's maximum acceleration in this process.

**51.** A book sits on a horizontal top of a car as the car accelerates horizontally from rest. If the static coefficient of friction between car top and book is 0.45, what is the maximum acceleration the car can have if the book is not to slip?

**52.** A 700-N man stands on a scale on the floor of an elevator. The scale records the force it exerts on whatever is on it. What is the scale reading if the elevator has an acceleration of

(A) 1.8 m/s$^2$ up?
(B) 1.8 m/s$^2$ down?
(C) 9.8 m/s$^2$ down?

## Friction; Inclined Planes

**53.** A 20-kg wagon is pulled along the level ground by a rope inclined at 30° above the horizontal. A friction force of 30 N opposes the motion. How large is the pulling force if the wagon is moving at

(A) constant speed?
(B) an acceleration of 0.40 m/s$^2$?

**54.** As shown in Figure 3.1, a force of 400 N pushes on a 25-kg box. Starting from rest, the box achieves a velocity of 2.0 m/s in a time of 4 s. Find the coefficient of sliding friction between box and floor.

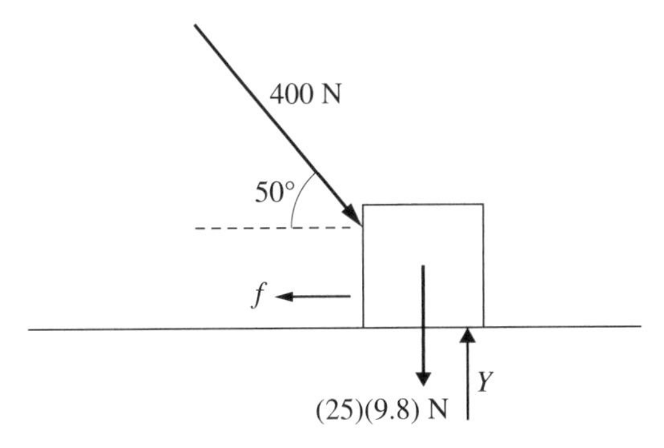


**Figure 3.1**

**55.** A 12-kg box is released from the top of an incline that is 5.0 m long and makes an angle of 40° to the horizontal. A 60-N friction force impedes the motion of the box.

(A) What will be the acceleration of the box?
(B) How long will it take to reach the bottom of the incline?

**56.** An inclined plane makes an angle of 30° with the horizontal. Neglecting friction forces, find the constant force, applied parallel to the plane, required to cause a 15-kg box to speed up, moving

(A) up the plane with acceleration 1.2 m/s$^2$
(B) down the incline with acceleration 1.2 m/s$^2$

**57.** A 400-g block originally moving at 120 cm/s coasts 70 cm along a tabletop before coming to rest. What is the coefficient of friction between block and table?

**58.** How large a force parallel to a 30° incline will allow a 5.0 kg box to speed up in a direction up the incline with an acceleration of 0.20 m/s$^2$

(A) if friction is negligible
(B) if the coefficient of friction is 0.30

**59.** An 8.0-kg box is released on a 30° incline and accelerates down the incline at 0.30 m/s$^2$. Find the frictional force impeding its motion. How large is the coefficient of friction in this situation?

## Two-Object and Other Problems

**60.** In Figure 3.2, find the acceleration of the cart that is required to prevent block $B$ from falling. The coefficient of static friction between the block and the cart is $\mu_s$.

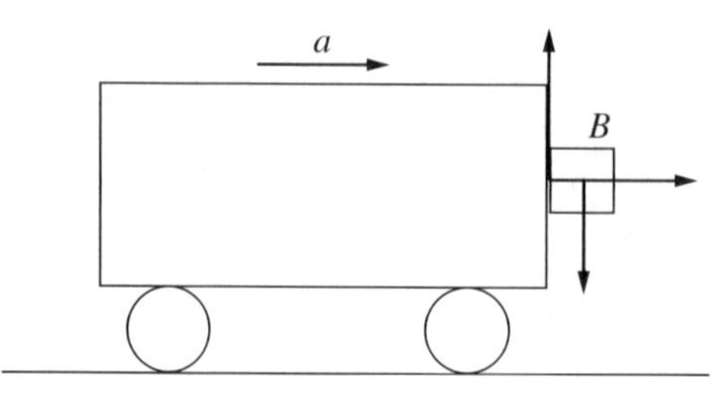


**Figure 3.2**

**61.** In Figure 3.3, mass $A$ is 15 kg and mass $B$ is 11 kg. If they are given an upward acceleration of 3 m/s$^2$ by pulling up on $A$, find the tensions $T_1$ and $T_2$.

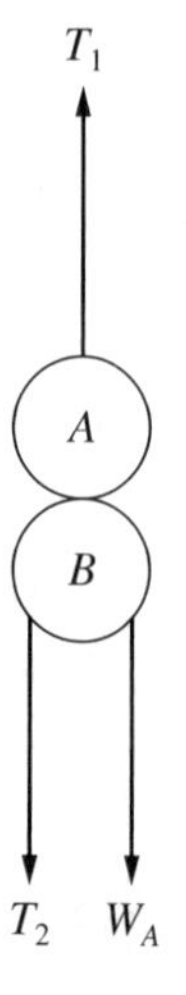


**Figure 3.3**

**62.** In Figure 3.4, if $F = 20$ N, $m_1 = m_2 = 3$ kg, and the acceleration is 0.50 m/s$^2$, what will be the tension in the connecting cord? How large is the frictional force on either block?

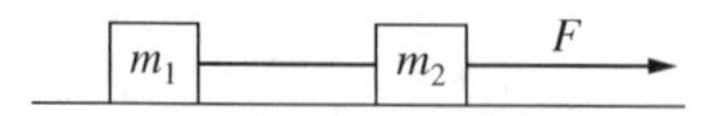


**Figure 3.4**

**63.** The device diagramed in Figure 3.5 is called an *Atwood's machine.* Let $m_1 = 2$ kg and $m_2 = 4$ kg. Assume the pulley to be frictionless and massless.

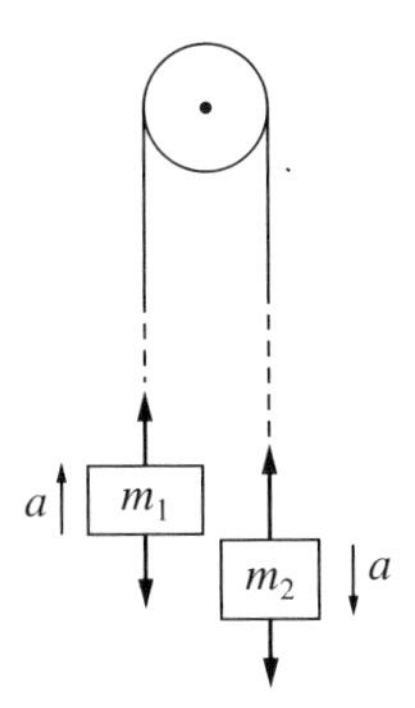


**Figure 3.5**

(A) How far will $m_2$ fall in time $t = 0.5$ s after the system is released?
(B) What is the tension in the light cord that connects the two masses?

**64.** An inclined plane making an angle of 25° with the horizontal has a pulley at its top. A 30-kg block on the plane is connected to a freely hanging 20-kg block by means of a cord passing over the pulley. Compute the distance the 20-kg block will fall in 2 s, starting from rest. Neglect friction.

**65.** Three blocks with masses 6 kg, 9 kg, and 10 kg are connected as shown in Figure 3.6. The coefficient of kinetic friction between the table and the 10-kg block is 0.2. Find

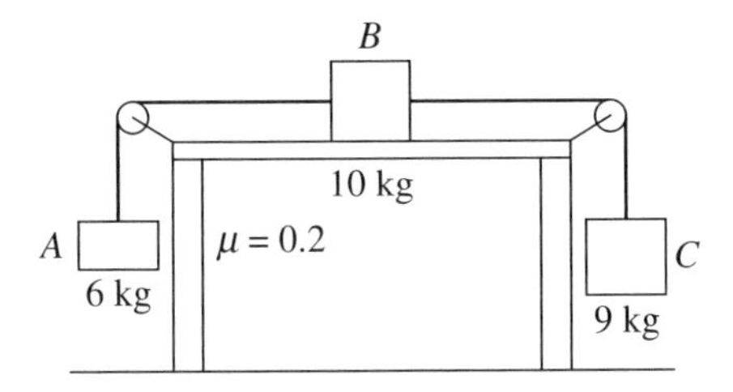


**Figure 3.6**

(A) the acceleration of the system as it moves to the right
(B) the tensions in the cord on the left and in the cord on the right

**66.** The cords holding the two masses shown in Figure 3.7 will break if the tension exceeds 15.0 N.

(A) What is the maximum upward acceleration one can give the masses without the cord breaking?
(B) Repeat if the strength is only 7.0 N.

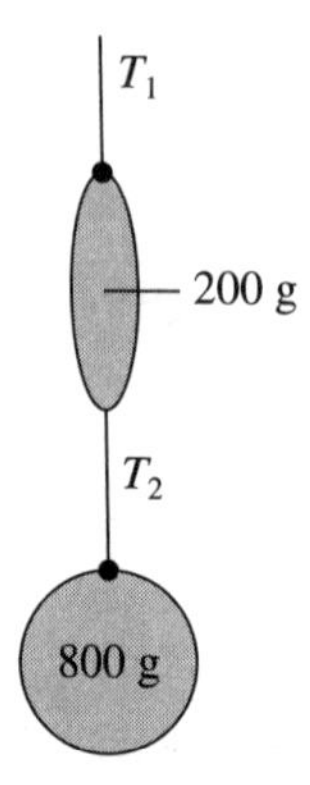


**Figure 3.7**

**67.** A 6.0-kg block rests on a smooth, frictionless table. A string attached to the block passes over a frictionless pulley, and a 3.0-kg mass hangs from the string, as shown in Figure 3.8.

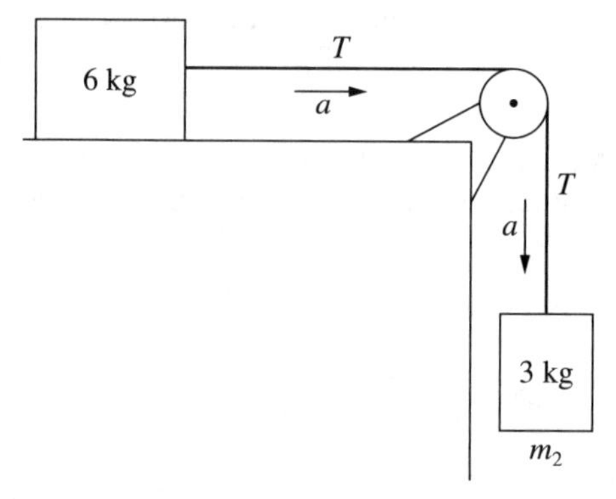


**Figure 3.8**

(A) What is the acceleration $a$?
(B) What is the tension $T$ in the string?

**68.** Suppose that blocks $A$ and $B$ have masses of 2 and 6 kg, respectively, and are in contact on a smooth horizontal surface. If a horizontal force of 6 N pushes them, calculate

(A) the acceleration of the two-object system
(B) the force that the 2-kg block exerts on the other block

CHAPTER 4

# Motion in a Plane

**69.** A boat can travel at a speed of 8 km/h in still water on a lake. In the flowing water of a stream, it can move at 8 km/h relative to the water in the stream. If the stream speed is 3 km/h, how fast can the boat move past a tree on the shore in traveling

(A) upstream
(B) downstream

**70.** A plane is traveling eastward at an airspeed of 500 km/h. But a 90 km/h wind is blowing southward. What are the direction and speed of the plane relative to the ground?

**71.** A marble with speed 20 cm/s rolls off the edge of a table 80 cm high. How long does it take to drop to the floor? How far, horizontally, from the table edge does the marble strike the floor?

**72.** In an ordinary television set, the electron beam consists of electrons shot horizontally at the television screen with a speed of about $5 \times 10^7$ m/s. How far does a typical electron fall as it moves the approximately 40 cm from the electron gun to the screen?

**73.** A body projected upward from the level ground at an angle of 50° with the horizontal has an initial speed of 40 m/s. How much time will elapse before it hits the ground?

**74.** In Problem 73, how far from the starting point will the body hit the ground, and at what angle with the horizontal?

**75.** A cannonball is projected downward at an angle of 30° with the horizontal from the top of a building 170 m high. Its initial speed is 40 m/s. Neglecting air resistance, how much time would elapse before it hits the ground?

**76.** In Question 75, find out how far from the foot of the building the cannonball will strike and at what angle with the horizontal.

**77.** A hose lying on the ground shoots a stream of water upward at an angle of 40° to the horizontal. The speed of the water is 20 m/s as it leaves the hose. How high up will it strike a wall that is 8 m away?

**78.** A projectile is fired with initial velocity $v_0 = 95$ m/s at an angle $\theta = 50°$. After 5 s it strikes the top of a hill. What is the elevation of the hill above the point of firing? At what horizontal distance from the gun does the projectile land?

**79.** A stunt flier is moving at 15 m/s parallel to the flat ground 100 m below. How large must the horizontal distance $x$ from plane to target be if a sack of flour released from the plane is to strike the target?

**80.** A cart is moving horizontally along a straight line with constant speed 30 m/s. A projectile is to be fired from the moving cart in such a way that it will return to the cart after the cart has moved 80 m. At what speed and at what angle, both relative to the cart, must the projectile be fired?

**81.** A ball is thrown upward from the top of a 35-m tower, as in Figure 4.1, with initial velocity $v_0 = 80$ m/s at an angle $\theta = 25°$.

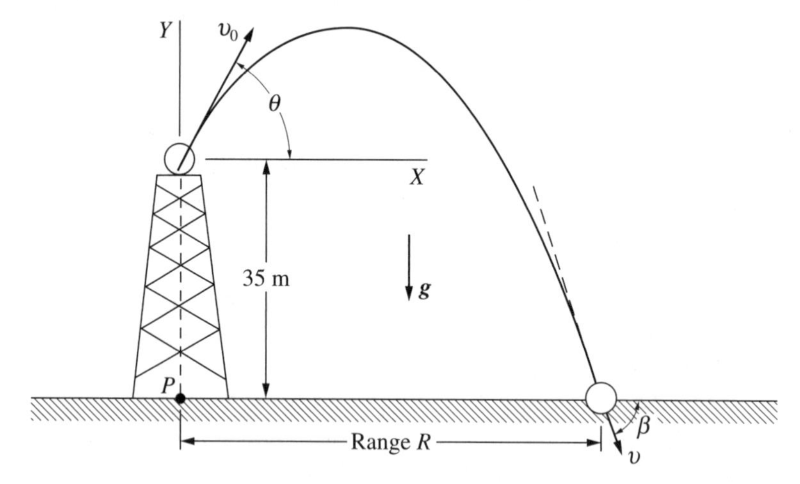


**Figure 4.1**

(A) Find the time to reach the ground and the distance $R$ from $P$ to the point of impact.
(B) Find the magnitude and direction of the velocity at the moment of impact.

**82.** A small ball is fastened to a string 24 cm long and suspended from a fixed point $P$ to make a conical pendulum, as shown in Figure 4.2. The ball describes a horizontal circle about a center vertically under point $P$, and the string makes an angle of 15° with the vertical. Find the speed of the ball.

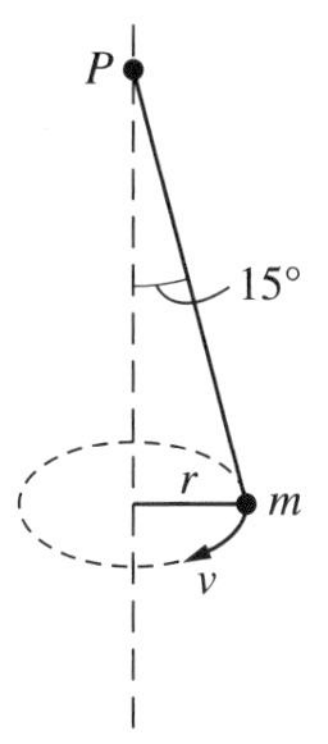


**Figure 4.2**

**83.** Find the maximum speed with which an automobile can round a curve of 80-m radius without slipping if the road is unbanked and the coefficient of friction between the road and the tires is 0.81.

**84.** The moon's radius is $1.74 \times 10^6$ m, and the acceleration due to gravity on the moon is 1.63 m/s$^2$. For a spaceship orbiting the moon just above its surface, find

(A) its speed
(B) its orbital period

**85.** While driving around a curve of 200-m radius, an engineer notes that a pendulum in the car hangs at an angle of 15° to the vertical. What should the speedometer read?

**86.** A 180-lb pilot is executing a vertical loop of radius 2000 ft at 350 mi/h. With what force does the seat press upward against him at the bottom of the loop?

**87.** The designer of a roller coaster wishes the riders to experience "weightlessness" as they round the top of one hill. How fast must the car be going if the radius of curvature at the hilltop is 20 m?

**88.** A huge pendulum consists of a 200-kg ball at the end of a cable 15 m long. If the pendulum is drawn back to an angle of 37° and released, what maximum force must the cable withstand as the pendulum swings back and forth?

## Law of Universal Gravitation; Satellite Motion

**89.** Two 16-lb, 7.3 kg, shot spheres (as used in track meets) are held 2 ft apart. What is the force of attraction between them?

**90.** The earth's radius is about 6370 km. An object that has a mass of 20 kg is taken to a height of 160 km above the earth's surface.

(A) What is the object's mass at this height?

(B) How much does the object weigh (i.e., how large a gravitational force does it experience) at this height?

**91.** The radius of the earth is about 6370 km, while that of Mars is about 3440 km. If an object weighs 200 N on earth, what would it weigh, and what would be the acceleration due to gravity on Mars? Mars has a mass 0.11 that of earth.

**92.** The moon orbits the earth in an approximately circular path of radius $3.8 \times 10^8$ m. It takes about 27 days to complete one orbit. What is the mass of the earth as obtained from these data?

**93.** The sun's mass is about $3.2 \times 10^5$ times the earth's mass. The sun is about 400 times as far from the earth as the moon is. What is the ratio of the magnitude of the pull of the sun on the moon to that of the pull of the earth on the moon? (It may be assumed that the sun-moon distance is constant and equal to the sun-earth distance.)

**94.** Communication satellites are placed in orbit above the equator in such a way that they remain stationary above a given point on earth below. How high above the surface of the earth is such a synchronous orbit? ($R_e = 6400$ km, $M_e = 5.98 \times 10^{24}$ kg)

CHAPTER 5

# Work and Energy

## Work Done by a Force

**95.** Figure 5.1 shows an overhead view of two horizontal forces pulling a box along the floor.

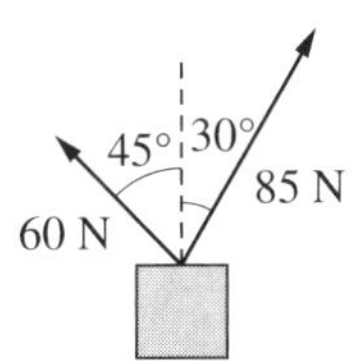


**Figure 5.1**

(A) How much work does each force do as the box is displaced 70 cm along the broken line?
(B) What is the total work done by the two forces in pulling the box this distance?

**96.** A horizontal force $F$ pulls a 20-kg carton across the floor at constant speed. If the coefficient of sliding friction between carton and floor is 0.60, how much work does $F$ do in moving the carton 3.0 m?

**97.** The coefficient of kinetic friction between a 20-kg box and the floor is 0.40. A pulling force is directed 37° above the horizontal. How much work does the pulling force do on the box in pulling it 8.0 m across the floor at constant speed?

**98.** Repeat Question 97 if the force pushes rather than pulls on the box and is directed 37° below horizontal.

**99.** How much work is done against gravity in lifting a 3-kg object through a distance of 40 cm?

**100.** A 4-kg object is slowly lifted 1.5 m.

(A) How much work is done against gravity?

(B) Repeat if the object is lowered instead of lifted.

**101.** If the $x$-directed force exerted on a cart by a boy varies with position, as shown in Figure 5.2, how much work does the boy do on the cart?

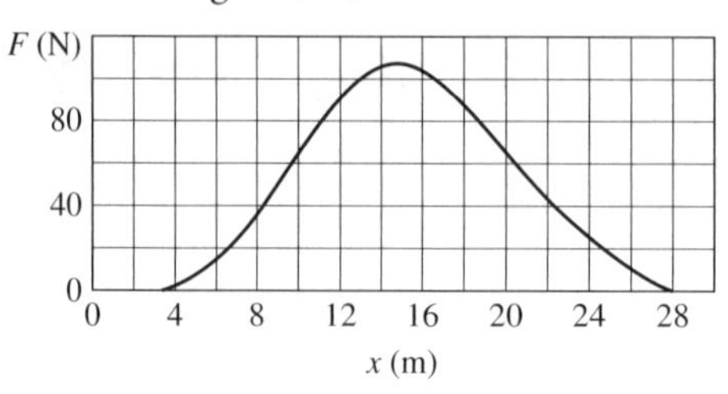


**Figure 5.2**

**102.** A 1200-kg car going 30 m/s applies its brakes and skids to rest. If the friction force between the sliding tires and the pavement is 6000 N, how far does the car skid before coming to rest?

**103.** A 200-kg cart is pushed slowly up an incline. How much work does the pushing force do in moving the object up along the incline to a platform 1.5 m above the starting point if friction is negligible?

**104.** Repeat Problem 103 if the distance along the incline to the platform is 7 m and a friction force of 150 N opposes the motion.

**105.** A rock weighing 20 N falls from a height of 16 m and sinks 0.6 m into the ground. From energy considerations, find the average force $f$ between the rock and the ground as the rock sinks. See Figure 5.3.

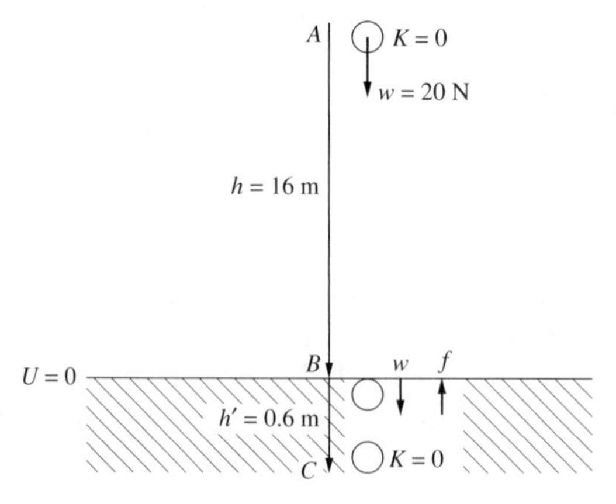


**Figure 5.3**

**106.** A delivery boy wishes to launch a 2.0-kg package up an inclined plane with sufficient speed to reach the top of the incline. The plane is 3.0 m long and is inclined at 20°. The coefficient of kinetic friction between the package and the plane is 0.40. What minimum initial kinetic energy must the boy supply to the package?

**107.** (A) If the simple pendulum shown in Figure 5.4 is released from point *A*, what will be the speed of the ball as it passes through point *C*?
(B) What is the speed of the ball at point *B*?

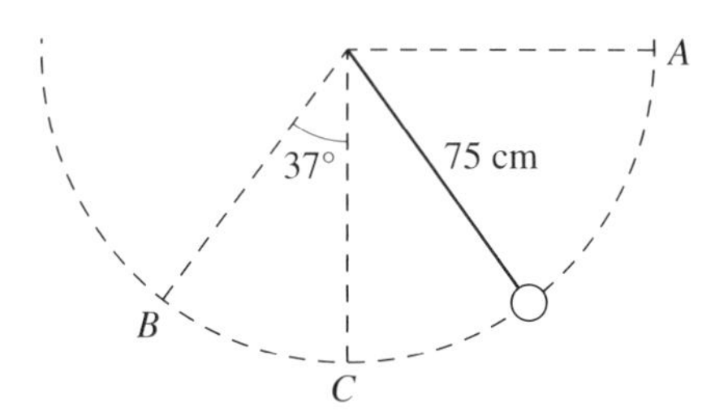


**Figure 5.4**

**108.** Figure 5.5 shows the plan for a proposed roller coaster track. Each car will start from rest at point *A* and will roll with negligible friction. It is important that there be at least some small normal force exerted by the track on the car at all points; otherwise, the car would leave the track. What is the minimum safe value for the radius of curvature at point *B*?

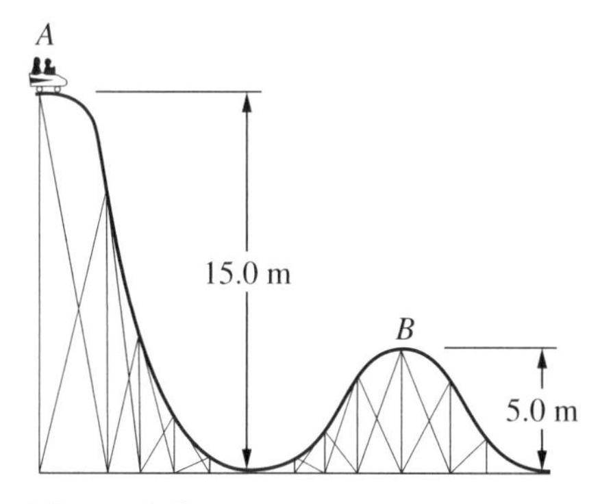


**Figure 5.5**

**109.** When a 300-g mass is hung from the end of a vertical spring, the spring's length is 40 cm. With 500 g hanging from it, its length is 50 cm. What is the spring constant of the spring?

**110.** A spring that stretches 10 cm under a load of 200 g requires how much work to stretch it 5 cm from its equilibrium position? How much work is required to stretch it the next 5 cm?

**111.** A block falls from a table 0.6 m high. It lands on an ideal, massless, vertical spring with a force constant of 2.4 kN/m. The spring is initially 25 cm high, but it is compressed to a minimum height of 10 cm before the block is stopped. Find the mass of the block.

CHAPTER 6

# Power, Impulse, and Momentum

## Power

**112.** The 4 metric ton (4000-kg) hammer of a pile driver is lifted 1.0 m in 2.0 s. What power does the engine furnish to the hammer? Assume that there is no acceleration of the hammer while it is being lifted.

**113.** A 1000-kg auto travels up a 3 percent grade at 20 m/s. Find the horsepower required, neglecting friction. (Note that a horsepower is defined as 756 watts.)

**114.** In unloading grain from the hold of a ship, an elevator lifts the grain through a distance of 12 m. Grain is discharged at the top of the elevator at a rate of 2.0 kg each second, and the discharge speed of each grain particle is 3.0 m/s. Find the minimum-power motor that can elevate grain in this way.

## Impulse-Momentum

**115.** A 3.0-kg block slides on a frictionless horizontal surface, first moving to the left at 50 m/s. It collides with a spring as it moves left, compresses the spring, and is brought to rest momentarily. The block continues to be accelerated to the right by the force of the compressed spring. Finally, the block moves to the right at 40 m/s. The block remains in contact with the spring for 0.020 s. What were the magnitude and direction of the impulse of the spring on the block? What was the spring's average force on the block?

## Inelastic Collisions

**116.** An 8-g bullet is fired horizontally into a 9-kg block of wood and sticks in it. The block, which is free to move, has a velocity of 40 cm/s after impact. Find the initial velocity of the bullet.

**117.** A 16-g mass is moving in the $+x$ direction at 30 cm/s while a 4-g mass is moving in the $-x$ direction at 50 cm/s. They collide head-on and stick together. Find their velocity after collision.

**118.** A 20-g bullet moving horizontally at 50 m/s strikes a 7-kg block resting on a table. The bullet embeds in the block after collision. Find

(A) the speed of the block after collision
(B) the frictional force between the table and block if the block moves 1.5 m before stopping

**119.** A 1.0-kg steel ball 4.0 m above the floor is released, falls, strikes the floor, and rises to a maximum height of 2.5 m. Find the momentum transferred from the ball to the floor in the collision.

**120.** A mass $A$ of 0.8 kg moving to the right with a speed of 5 m/s collides head-on with a mass $B$ of 1.2 kg moving in the opposite direction with a speed of 4 m/s. After the collision, $A$ is moving to the left with a speed of 4 m/s. Find the velocity of $B$ after collision and the coefficient of restitution.

**121.** As shown in Figure 6.1, a 15-g bullet is fired horizontally into a 3-kg block of wood suspended by a long cord. The bullet sticks in the block. Compute the velocity of the bullet if the impact causes the block to swing 10 cm above its initial level.

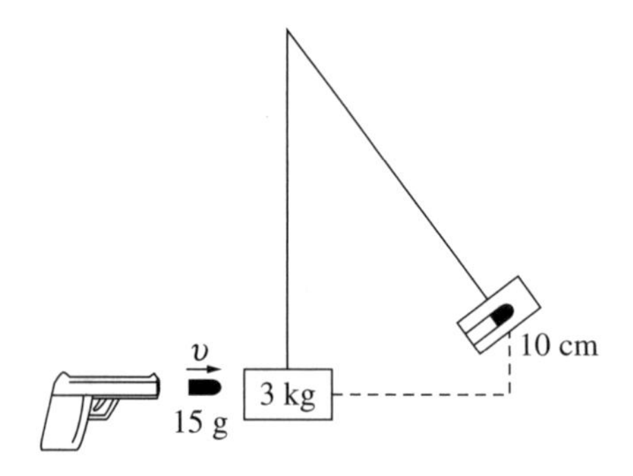


**Figure 6.1**

**122.** A 5.00-g bullet traveling 250 m/s strikes and embeds itself in a 2.495-kg block held on a frictionless table by a spring with constant $k = 40$ N/m. Find the speed of block and bullet immediately after the collision and the distance the spring is compressed.

**123.** A 2.0-kg block rests over a small hole on a table. A 15.0-g bullet is shot from below through the hole into the block, where it lodges. How fast was the bullet going if the block rises 1.30 m above the table?

**124.** A 1200-kg car is moving east at 30.0 m/s and collides with a 3600-kg truck moving at 20.0 m/s in a direction 60° north of east. The vehicles interlock and move off together. Find their common velocity.

**125.** The nucleus of a certain atom has a mass of $3.8 \times 10^{-25}$ kg and is at rest. The nucleus is radioactive and suddenly ejects from itself a particle of mass $6.6 \times 10^{-27}$ kg and speed $1.5 \times 10^{7}$ m/s. Find the recoil speed of the nucleus left behind.

**126.** Two blocks of masses 200 g and 500 g sit on a frictionless table with an essentially massless spring placed between them. They are pushed together until an energy of 3.0 J is stored in the spring. When released, the masses shoot off in opposite directions. What is the speed of

(A) the center of mass?
(B) the 500-g block?

## Equilibrium of Rigid Bodies

**127.** A uniform beam weighs 200 N and holds a 450-N weight, as shown in Figure 6.2. Find the magnitudes of the forces exerted on the beam by the two supports at its ends.

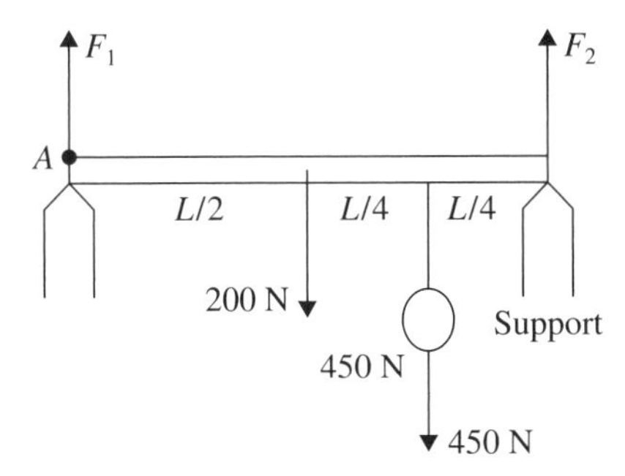


**Figure 6.2**

**128.** A uniform 100-N pipe is used as a lever, as shown in Figure 6.3. Where must the fulcrum be placed if a 500-N weight at one end is to balance a 200-N weight at the other end? How much load must the support hold?

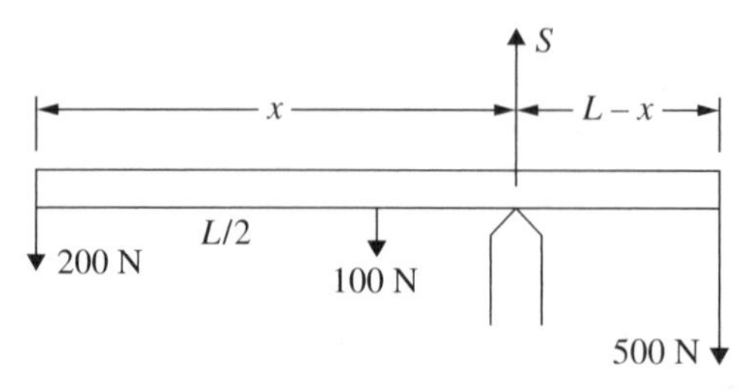


**Figure 6.3**

**129.** Where must an 800-N weight be hung on a uniform 100-N pole so that a boy at one end supports one-third as much as a man at the other end?

**130.** A uniform 200-N board of length $L$ has two weights hanging from it, 300 N at $L/3$ from one end and 400 N at $3L/4$ from the same end (Figure 6.4). What single additional force acting on the board will produce equilibrium?

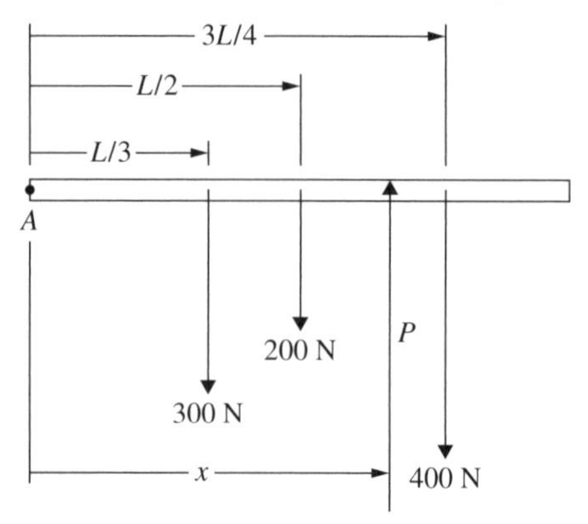


**Figure 6.4**

**131.** In reference to Figure 6.5, the plank is uniform and weighs 500 N. How large must $W$ be if $T_1$ and $T_2$ are to be equal?

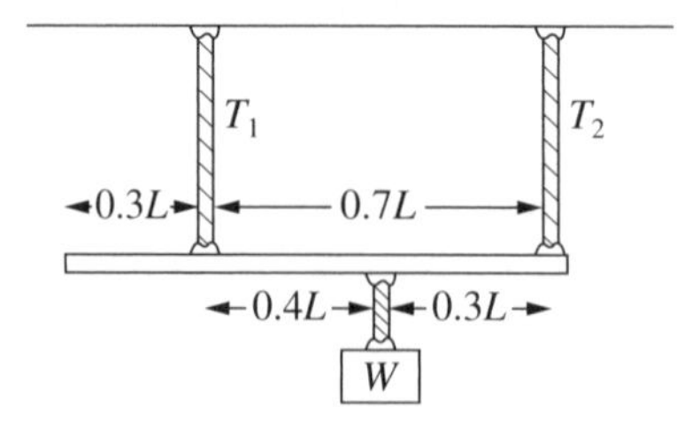


**Figure 6.5**

CHAPTER 7

# Rotational Motion

## Rotational Kinematics

**132.** A flywheel originally at rest is to reach an angular velocity of 36 rad/s in 6.0 s.

(A) What constant angular acceleration must it have?
(B) What total angle does it turn through in the 6.0 s?

**133.** A wheel turning with angular speed of 30 rev/s is brought to rest with a constant acceleration. It turns 60 rev before it stops.

(A) What is its angular acceleration?
(B) What time elapses before it stops?

**134.** A belt runs on a wheel of 30-cm radius. During the time that the wheel coasts uniformly to rest from an initial speed of 2.0 rev/s, 25 m of belt length passes over the wheel. Find the acceleration of the wheel and the number of revolutions it turns while stopping.

**135.** A car accelerates uniformly from rest to a speed of 15 m/s in a time of 20 s. Find the angular acceleration of one of its wheels and the number of revolutions turned by a wheel in the process. The radius of the car wheel is 0.33 m.

## Torque and Rotation

**136.** A grindstone has a moment of inertia of $1.6 \times 10^{-3}$ kg · m$^2$. When a constant torque is applied, the flywheel reaches an angular velocity of 1200 rev/min in 15 s. Assuming it started from rest, find

(A) the angular acceleration
(B) the unbalanced torque applied
(C) the angle turned through in the 15 s
(D) the work *W* done on the flywheel by the torque

**137.** Three children are sitting on a seesaw in such a way that it balances. A 20- and a 30-kg boy are on opposite sides at a distance of 2.0 m from the pivot. If the third boy jumps off, thereby destroying the balance, what is the initial angular acceleration of the board? (Neglect the weight of the board.)

**138.** A pendulum consists of a small mass $m$ at the end of a string of length $L$. The pendulum is pulled aside to an angle $\theta$ with the vertical and released. At the instant of release, using the suspension point as axis, find

(A) the torque on the pendulum

(B) its angular acceleration

## Moment of Inertia

**139.** Two thin hoops of masses $m_1$ and $m_2$ have radii $a_1$ and $a_2$, respectively. They are mounted rigidly on a frame of negligible mass. Find the system's moment of inertia about an axis through the center and perpendicular to the page. How large a torque must be applied to the system to give it an angular acceleration $\alpha$ about this axis, provided it is free to turn? Repeat for the axis $AA'$.

**140.** Three thin uniform rods each of mass $M$ and length $L$ lie along the $x$, $y$, and $z$ axes, with one end of each at the origin. Given $I = \frac{1}{12} ML^2$ for a rod pivoted about its center of mass, find $I$ about the $z$ axis for the three-rod system.

**141.** A rod of length $L$ is composed of a uniform length $\frac{1}{2}L$ of wood whose mass is $m_w$ and a uniform length $\frac{1}{2}L$ of brass whose mass is $m_b$.

(A) Find $I$ for the rod about an axis perpendicular to the rod and through its center.

(B) Repeat for a parallel axis through the wood end.

**142.** As shown in Figure 7.1, a girl on a rotating platform holds a pendulum in her hand. The pendulum is at a radius of 6.0 m from the center of the platform. The rotational speed of the platform is 0.020 rev/s. It is found that the pendulum hangs at an angle $\theta$ to the vertical, as shown. Find $\theta$.

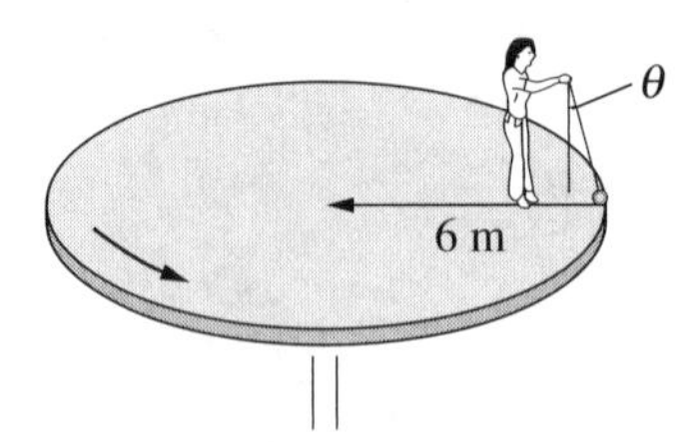


**Figure 7.1**

**143.** The rotational speed of the earth is 1 rev/day, or $1.16 \times 10^{-5}$ rev/s, and the earth's radius is $6.37 \times 10^6$ m. If a man at the equator is standing on a spring scale, by what percent would his apparent weight increase if the earth were to stop rotating? A man at the north pole?

**144.** A 20-mg bug sits on the smooth edge of a 25-cm-radius phonograph record as the record is brought up to its normal rotational speed of 45 rev/min. How large must the coefficient of friction between the bug and record be if the bug is not to slip off?

## Problems Involving Cords Around Cylinders, Rolling Objects, and So On

**145.** A 25-kg wheel has a radius of 40 cm and turns freely on a horizontal axis. The radius of the wheel is 30 cm. A 1.2-kg mass hangs at the end of a cord that is wound around the rim of the wheel. This mass falls and causes the wheel of moment of inertia $\frac{1}{2}MR^2$ to rotate. Find the acceleration of the falling mass and the tension in the cord.

**146.** (A) The rope shown in Figure 7.2 is wound around a cylinder of mass 4.0 kg and $I = 0.020$ kg · m², about the cylinder axis. If the cylinder rolls without slipping, what is the linear acceleration of its center of mass? What is the frictional force? Use an axis along the cylinder axis for your computation.

(B) What happens if the frictional force between table and cylinder is negligible? Choose any axis for your computation.

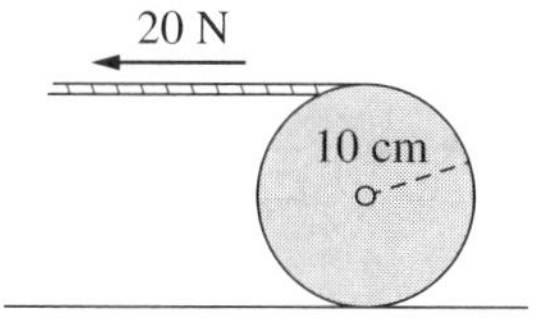


**Figure 7.2**

**147.** The moment of inertia of the wheel in Figure 7.3 is 8.0 kg · m². Its radius is 40 cm. Find the angular acceleration of the wheel caused by the 10.0-kg mass if the frictional force between the mass and the incline is 30 N.

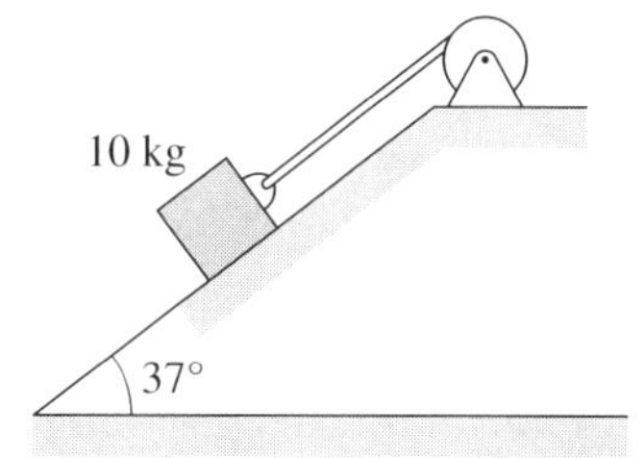


**Figure 7.3**

## Energy and Power

**148.** A flywheel having a moment of inertia of 900 kg · m$^2$ rotates at a speed of 120 rev/min. It is slowed down by a brake to a speed of 90 rev/min. How much work was done by the brake?

**149.** The rigid rod joining the three masses in Figure 7.4 has negligible mass. It is pivoted at one end so that it can swing in a vertical plane. If it is released from the position shown, how fast will the bottom mass be moving when the rod is vertical?

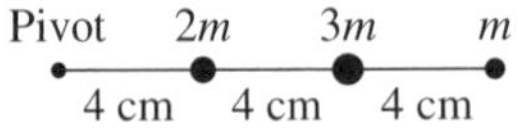

Pivot $2m$ $3m$ $m$

4 cm 4 cm 4 cm

**Figure 7.4**

**150.** As shown in Figure 7.5, a uniform solid sphere of moment inertia $I = \frac{2}{5}$ MR$^2$ rolls on a horizontal surface at 20 m/s. It then rolls up the incline shown. If friction losses are negligible, what will be the value of $h$ where the ball stops?

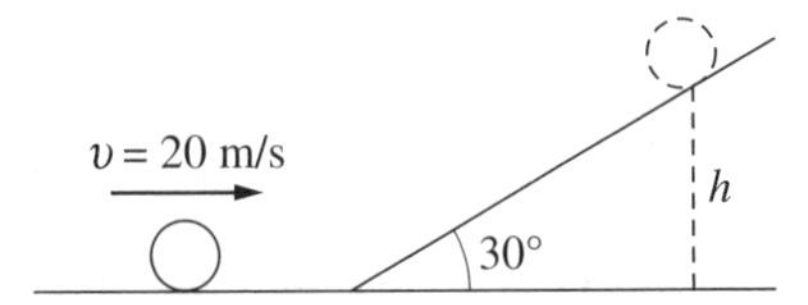


**Figure 7.5**

**151.** As a solid disk rolls over the top of a hill on a track, its speed is 80 cm/s. If friction losses are negligible, how fast is the disk moving when it is 18 cm below the top?

**152.** A cord 3 m long is coiled around the axle of a wheel. The cord is pulled with a constant force of 40 N. When the cord leaves the axle, the wheel is rotating at 2 rev/s. Determine the moment of inertia of the wheel and axle. Neglect friction.

**153.** A cylinder of radius 20 cm is mounted on an axle coincident with its axis so as to be free to rotate. A cord is wound on it, and a 50-g mass is hung from it. If, after being released, the mass drops 100 cm in 12 s, find the moment of inertia of the cylinder.

**154.** It is proposed to use a uniform disk 50 cm in radius turning at 300 rev/s as an energy-storage device in a bus. How much mass must the disk have if it is to be capable, while coasting to rest, of furnishing the energy equivalent of a 100-hp (75 kW) motor operating for 10 min?

## Angular Impulse; The Physical Pendulum

**155.** A uniform rod is pivoted at its end, as shown in Figure 7.6.

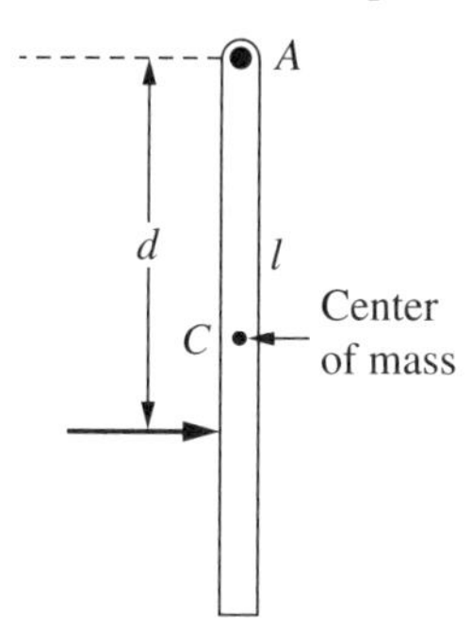


**Figure 7.6**

(A) What is the period of oscillation of the rod when it is suspended from $A$?
(B) What is the length of the simple pendulum having the same period?

**156.** A ring of mass $M$ and radius $R$ is hung from a knife edge so that the ring can swing in its own plane as a physical pendulum. Find the period $T_1$ of small oscillations.

**157.** An ice skater spins with arms outstretched at 1.9 rev/s. Her moment of inertia at this time is 1.33 kg · m$^2$. She pulls in her arms to increase her rate of spin. If her moment of inertia is 0.48 kg · m$^2$ after she pulls in her arms, what is her new rate of rotation?

**158.** Consider a satellite orbiting earth, as shown in Figure 7.7. Find the ratio of its speed at perihelion to that at aphelion.

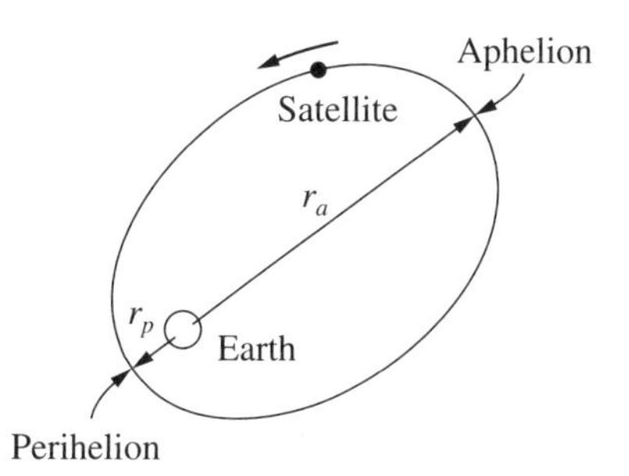


**Figure 7.7**

**159.** Assume that the center of mass of a girl crouching in a light swing has been raised to 1.2 m (see Figure 7.8). The girl weighs 400 N, and her center of mass is 3.7 m from the pivot of the swing while she is in the crouched position. The swing is released from rest, and at the bottom of the arc the girl stands up instantaneously, thus raising her center of mass 0.6 m (returning it to its original level). Find the height of her center of mass at the top of the arc.

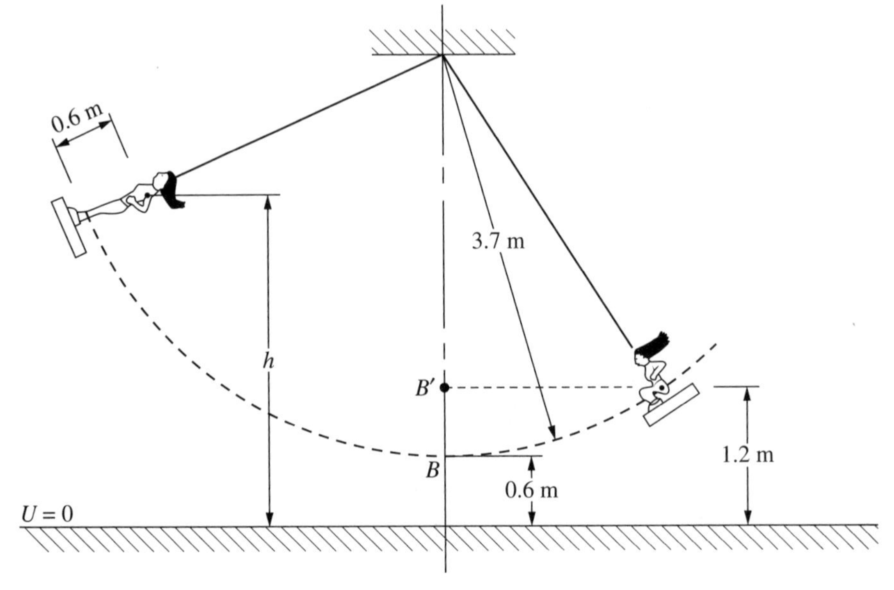


**Figure 7.8**

**160.** As shown in Figure 7.9, sand drops onto a disk rotating freely about an axis. The moment of inertia of the disk about this axis is $I$, and its original rotational rate was $\omega_0$. What is its rate of rotation after a mass $M$ of sand has accumulated on the disk at radius $b$?

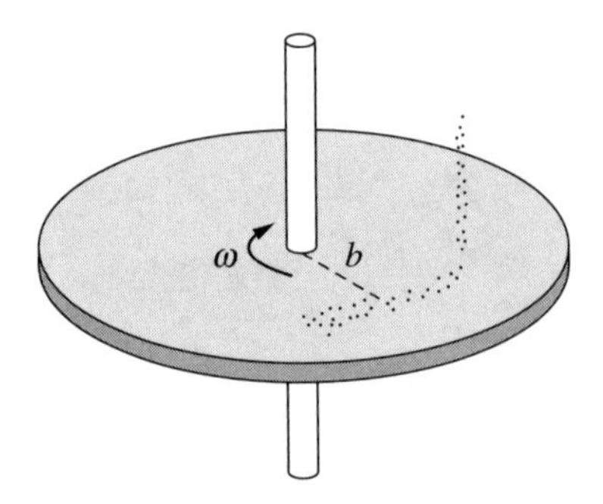


**Figure 7.9**

CHAPTER 8

# Simple Harmonic Motion

## Oscillations of a Mass on a Spring

**161.** A 50-g mass hangs at the end of a Hookean spring. When 20 g more are added to the end of the spring, it stretches 7.0 cm more.

(A) Find the spring constant.
(B) If the 20 g are now removed, what will be the period of the motion?

**162.** A body of weight 27 N hangs on a long spring of such stiffness that an extra force of 9 N stretches the spring 0.05 m. If the body is pulled downward and released, what is its period?

**163.** A 0.5-kg body performs simple harmonic motion with a frequency of 2 Hz and an amplitude of 8 mm. Find the maximum velocity of the body, its maximum acceleration, and the maximum restoring force to which the body is subjected.

**164.** A mass of 250 g hangs on a spring and oscillates vertically with a period of 1.1s. To double the period, what mass must be added to the 250 g? (Ignore the mass of the spring.)

**165.** In a certain engine, a piston undergoes vertical SHM with amplitude 7 cm. A washer rests on top of the piston. As the motor is slowly speeded up, at what frequency will the washer no longer stay in contact with the piston?

**166.** A block of mass 4 kg hangs from a spring of force constant $k = 400$ N/m. The block is pulled down 15 cm below equilibrium and released.

(A) Find the amplitude, frequency, and period of the motion.
(B) Find the kinetic energy when the block is 10 cm above equilibrium.

**167.** A certain pendulum clock keeps good time on the earth. If the same clock were placed on the moon, where objects weigh only one-sixth as much as on earth, how many "seconds" will the clock tick out in an actual time of 1 min?

**168.** Find the frequency of oscillation when a meterstick is hung as a compound pendulum with the pivot at its 90-cm mark. Repeat for the pivot at the 50.1-cm mark.

**169.** A motor vehicle to carry astronauts on the surface of the moon has a spring suspension and has a natural up-and-down frequency of 0.40 Hz when fully loaded on the earth. Find its natural frequency on the moon, where it and everything in it will weigh only about one-sixth what it weighs on earth.

CHAPTER 9

# Hydrostatics

## Pressure and Density

**170.** The pressure gauge shown in Figure 9.1 has a spring for which $k = 60$ N/m, and the area of the piston is 0.50 $cm^2$. Its right end is connected to a closed container of gas at a gauge pressure of 30 kPa. Atmospheric pressure is 101 kPa. How far will the spring be compressed if the region containing the spring is

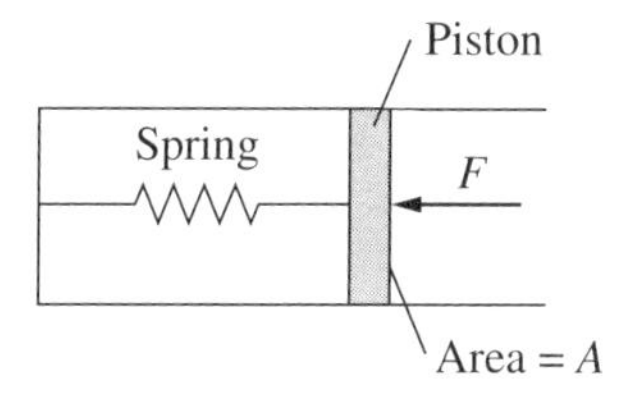


**Figure 9.1**

(A) in vacuum
(B) open to the atmosphere

**171.** A tank contains a pool of mercury 0.30 m deep, covered with a layer of water that is 1.2 m deep. The density of water is $1.0 \times 10^3$ kg/$m^3$, and that of mercury is $13.6 \times 10^3$ kg/$m^3$. Find the pressure exerted by the double layer of liquids at the bottom of the tank. Ignore the pressure of the atmosphere.

**172.** The manometer shown in Figure 9.2 uses mercury as its fluid. If atmospheric pressure is 100 kPa, what is the pressure of the gas in the container shown on the left?

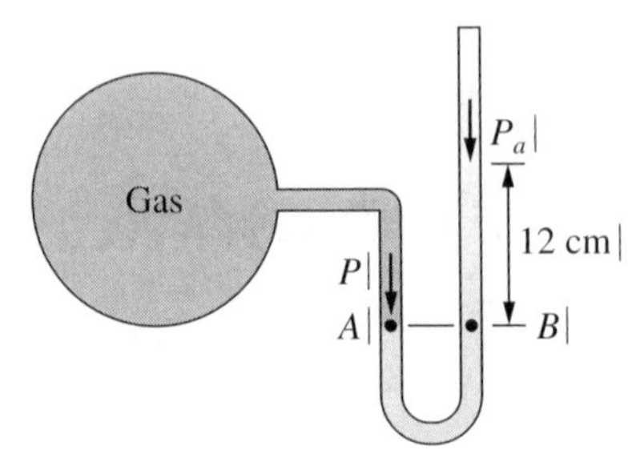


**Figure 9.2**

**173.** As shown in Figure 9.3, a weighted piston holds compressed gas in a tank. The piston and its weights have a mass of 20 kg. The cross-sectional area of the piston is 8 $cm^2$. What is the absolute pressure of the gas in the tank? What would a pressure gauge on the tank read?

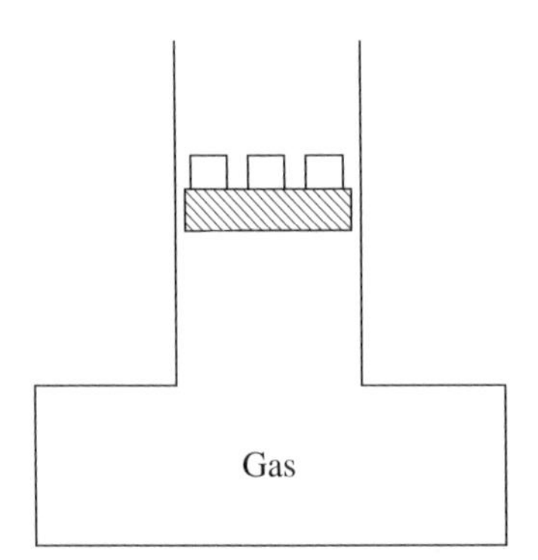


**Figure 9.3**

## Pascal's and Archimedes' Principles; Surface Tension

**174.** A hydraulic lift in a service station has a large piston 30 cm in diameter and a small piston 2 cm in diameter.

(A) What force is required on the small piston to lift a load of 1500 kg?
(B) What is the pressure increase due to the force in the confined liquid?

**175.** For the system shown in Figure 9.4, the cylinder on the left, at $L$, has a mass of 600 kg and a cross-sectional area of 800 cm$^2$. The piston on the right, at $S$, has cross-sectional area 25 cm$^2$ and negligible weight. If the apparatus is filled with oil ($\rho = 0.78$ g/cm$^3$), what is the force $F$ required to hold the system in equilibrium as shown?

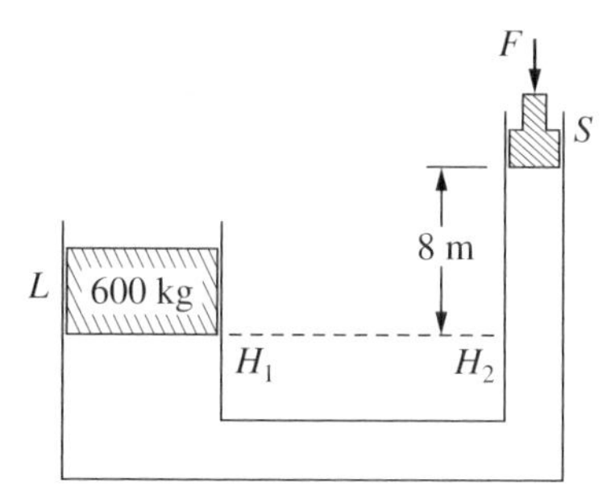


**Figure 9.4**

**176.** A block of wood weighing 71.2 N and of specific gravity 0.75 is tied by a string to the bottom of a tank of water in order to have the block totally immersed. What is the tension in the string?

**177.** A metal ball weighs 0.096 N. When suspended in water, it has an apparent weight of 0.071 N. Find the density of the metal.

**178.** The density of ice is 917 kg/m$^3$, and the approximate density of the seawater in which an iceberg floats is 1025 kg/m$^3$. What fraction of the iceberg is beneath the water surface?

**179.** What is the minimum volume of a block of wood (density = 850 kg/m$^3$) if it is to hold a 50-kg woman entirely above the water when she stands on it?

**180.** A beaker when partly filled with water has a total mass of 20.00 g. a piece of metal with density 3.00 g/cm$^3$ and volume 1.00 cm$^3$ is suspended by a thin string so that it is submerged in the water but does not rest on the bottom of the beaker. The beaker is then placed on a platform scale. What is the reading in the scale?

**181.** A solid cube of material is 0.75 cm on each edge. It floats in oil of density 800 kg/m$^3$ with one-third of the block out of the oil.

(A) What is the buoyant force on the cube?
(B) What is the density of the material of the cube?

**182.** A cubical copper block is 1.50 cm on each edge.

(A) What is the buoyant force on it when it is submerged in oil for which $\rho = 820\ \text{kg/m}^3$?

(B) What is the tension in the string that is supporting the block when submerged? $\rho_{Cu} = 8920\ \text{kg/m}^3$.

**183.** A balloon having a mass of 500 kg remains suspended motionless in the air. If the air density is $1.29\ \text{kg/m}^3$, what is the volume of the balloon in cubic meters?

CHAPTER 10

# Hydrodynamics

**184.** Consider the flow of a fluid at speed $v_0$ through a cylindrical pipe of radius $r$. What would be the speed of this fluid at a point where, because of a constriction in the pipe, the fluid is confined to a cylindrical opening of radius $r/4$?

**185.** Water is flowing smoothly through a closed-pipe system. At one point, the speed of the water is 3.0 m/s, while at another point 1.0 m higher, the speed is 4.0 m/s. If the pressure is 20 kPa at the lower point, what is the pressure at the upper point? What would the pressure at the upper point be if the water were to stop flowing and the pressure at the lower point were 18 kPa?

**186.** Water flows out of a pipe at the rate of 3.0 cm$^3$/s. Find the velocity of the water at a point in the pipe where its diameter is

(A) 0.50 cm
(B) 0.80 cm

**187.** In Figure 10.1, the system is filled to height $h$ with a liquid of density $\rho$. The atmospheric pressure is $p_{atm}$. Neglecting fluid friction, evaluate the pressure of the fluid at each of the points labeled 1, 2, 4, and 5. Compare the pressure at point 3 with that at point 5.

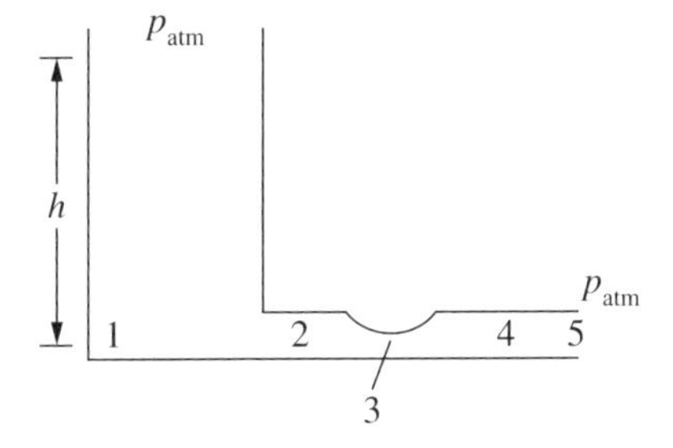


**Figure 10.1**

**188.** The tank shown in Figure 10.2 is kept filled with water to a depth of 8.0 m. Find the speed $v_b$ with which the jet of water emerges from the small pipe just at the bottom of the tank.

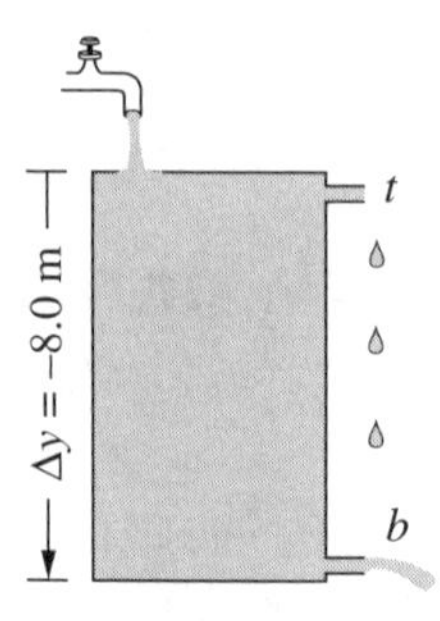


**Figure 10.2**

**189.** A water barrel stands on a table of height $h$. If a small hole is punched in the side of the barrel at its base, it is found that the resultant stream of water strikes the ground at a horizontal distance R from the barrel. What is the depth of water in the barrel?

**190.** A hole of area 1 $mm^2$ opens in the pipe near the lower end of a large water-storage tank, and a stream of water shoots from it. If the top of the water in the tank is 20 m above the point of the leak, how much water escapes in 1 s?

**191.** A water-filled can sits on a table. The water squirts out of a small hole in the side of the can, located a distance $y$ below the water surface. The height of the water in the can is $h$. At what distance $R$ from the base of the can, directly below the hole, does the water strike the table top? Neglect air resistance.

**192.** A hose shoots water straight up for a distance of 2.5 m. The end opening on the hose has an area of 0.75 $cm^2$. What is the speed of the water as it leaves the hose? How much water comes out in 1 min?

**193.** Water leaves a faucet with a downward velocity of 3.0 m/s. As the water falls below the faucet, it accelerates with acceleration g. The cross-sectional area of the water stream leaving the faucet is 1.0 $cm^2$. What is the cross-sectional area of the stream 0.50 m below the faucet?

**194.** A cylindrical tank 0.9 m in radius rests atop a platform 6 m high. Initially the tank is filled with water ($\rho = 1 \times 10^3$ kg/m$^3$) to a depth $h_0 = 3$ m. A plug whose area is 6.3 $cm^2$ is removed from an orifice in the side of the tank at the bottom. What is the speed of the stream as it strikes the ground?

CHAPTER 11

# Temperature and Thermal Expansion

## Temperature Scales; Linear Expansion

**195.** A rod 3 m long is found to have expanded 0.091 cm in length for a temperature rise of 60°C. What is the coefficient of linear expansion $\alpha$ for the material of the rod?

**196.** At 15°C, a bare wheel has a diameter of 30.000 in and the inside diameter of a steel rim is 29.930 in. To what temperature must the rim be heated so as to slip over the wheel?

**197.** Suppose that the standard of length was a 1-m-long bar of iron. What would be the maximum temperature variation of the bar if its length were to be preserved to an accuracy of 1 part per million?

**198.** An iron ball has a diameter of 6 cm and is 0.010 mm too large to pass through a hole in a brass plate when the ball and plate are at a temperature of 30°C. At what temperature, the same for ball and plate, will the ball just pass through the hole?

**199.** A steel rail 30 m long is firmly attached to the roadbed only at its ends. The sun raises the temperature of the rail by 50°C, causing the rail to buckle. Assuming that the buckled rail consists of two straight parts meeting in the center, calculate how much the center of the rail rises. The coefficient of linear expression for steel is $12 \times 10^{-6}\ k^{-1}$.

**200.** (A) An aluminum measuring rod, which is correct at 5°C, measures a certain distance as 88.42 cm at 35°C. Determine the error in measuring the distance due to the expansion of the rod.
(B) If this aluminum rod measures a length of steel as 88.42 cm at 35°C, what is the correct length of the steel at 35°C?

**201.** A steel tape measure is calibrated at 70°F. The width of a building lot is measured with the tape when the temperature is 15°F, and a reading of 150 ft is obtained. How great an error is produced by the temperature difference? The coefficient of linear expansion of steel is $(6.7 \times 10^{-6})$/°F.

**202.** A steel tape is calibrated at 20°C. On a cold day when the temperature is –15°C, what will be the percent error in the tape?

**203.** A grandfather clock has a pendulum made of brass. The clock is adjusted to have a period of 1 s exactly at 20°C. If operated at 30°C, how much will the clock be in error 1 week after it is set? Will it be fast or slow?

**204.** An open aluminum 300-mL container is full of glycerin at 20°C. What volume of glycerin overflows when the container is heated to 110°C? The coefficients of value expression are $77 \times 10^{-5}$ $k^{-1}$ for aluminum and $530 \times 10^{-6}$ $k^{-1}$ for glycerin.

**205.** The density of mercury at 0°C is 13,600 kg/m$^3$. Calculate the density of mercury at 50°C.

CHAPTER 12

# Heat and Calorimetry

## Heat and Energy; Mechanical Equivalent of Heat

**206.** Victoria Falls in Africa is 122 m in height. Calculate the rise in temperature of the water if all the potential energy lost in the fall is converted to heat.

**207.** A 2.2-g lead bullet is moving at 150 m/s when it strikes a bag of sand and is brought to rest. If all the frictional work is transferred to thermal energy in the bullet, what is the rise in temperature of the bullet as it is brought to rest?

**208.** Repeat if the bullet lodges in a 50-g block of wood that is free to move.

**209.** A lead bullet of mass $m$ is fired at a tree trunk and emerges on the other side. The speed of the bullet is 500 m/s as it enters and 300 m/s as it emerges. Assuming that 40 percent of the loss of kinetic energy is stored as heat in the bullet, calculate the rise in temperature of the bullet.

**210.** A mass $m$ of lead shot is placed at the bottom of a vertical cardboard cylinder that is 1.5 m long and closed at both ends. The cylinder is suddenly inverted so that the shot falls 1.5 m. By how much will the temperature of the shot increase if this process is repeated 100 times? Assume no heat loss.

**211.** A 60-kg boy running at 5.0 m/s while playing basketball falls to the floor and skids along on his leg until he stops. How many calories of heat are generated between his leg and the floor? Assume that all this heat energy is confined to a volume of 2.0 cm$^3$ of his flesh. What will be the temperature change of the flesh? Assume $c = 1.0$ cal/$g \cdot$ °C and $\rho =$ 950 kg/m$^3$ for flesh.

**212.** Cool water at 9.0°C enters a hot-water heater from which warm water at a temperature of 80°C is drawn at an average rate of 300 g/min. How much average electric power does the heater consume to provide hot water at this rate? Assume that there is negligible heat loss to the surroundings.

**213.** An electric heater supplies 1.8 kW of power in the form of heat to a tank of water. How long will it take to heat the 200 kg of water in the tank from 10 to 70°C? Assume heat losses to the surroundings to be negligible.

**214.** (A) A certain 6-g bullet melts at 300°C and has a specific heat capacity of 0.20 cal/g · °C and a heat of fusion of 15 cal/g. How much heat is needed to melt the bullet if it is originally at 0°C?

(B) Refer to part (A). What is the slowest speed at which the bullet can travel if it is to just melt when suddenly stopped?

**215.** A normal diet might furnish 2000 nutritionist's calories (2000 kcal) to a 60-kg person in a day. If this energy were used to heat the person with no losses to the surroundings, how much would the person's temperature increase? $c = 0.83$ cal/g · °C for the average person.

## Calorimetry, Specific Heats, Heats of Fusion, and Vaporization

**216.** A copper container of mass 0.30 kg contains 0.45 kg of water. Container and water are initially at room temperature, 20°C. A 1-kg block of metal is heated to 100°C and placed in the water in the calorimeter. The final temperature of the system is 40°C. Find the specific heat of the metal.

**217.** What will be the final temperature if 50 g of water at 0°C is added to 250 g of water at 90°C?

**218.** A student heated a 10-g iron nail for some time in a Bunsen-burner flame and then plunged the nail into 100 g of water at 10°C. The water temperature rose to 20°C. What was the temperature of the flame?

**219.** If 250 g of Ni at 120°C is dropped into 200 g of water at 10°C contained by a calorimeter of 20-cal/°C heat capacity, what will be the final temperature of the mixture?

**220.** How much water at 0°C is needed to cool 500 g of water at 80°C down to 20°C?

**221.** A 500-g piece of iron at 400°C is dropped into 800 g of oil at 20°C. If $c =$ 0.40 cal/g · °C for the oil, what will be the final temperature of the system? Assume no loss to the surroundings.

**222.** Initially, 48.0 g of ice at 0°C is in an aluminum calorimeter can of mass 2.0 g, also at 0°C. Then 75.0 g of water at 80°C is poured into the can. What is the final temperature?

**223.** How many grams of ice at 0°C must one add to a 200-g cup of coffee at 90°C to cool it to 60°C? Assume heat transfer with the surroundings to be negligible.

**224.** A calorimeter of mass 125 g contains 130 g of water at 20°C. A 6.1-g mass of steam at 100°C is introduced into the calorimeter and condensed into water. What is the final temperature of the water? Assume that no heat is lost to the surroundings and that the value of c for the calorimeter is 0.10 kcal/kg · °C.

# CHAPTER 13

# Heat Transfer

## Conduction

**225.** A slab having a thickness of 4 cm and measuring 25 cm on a side has a 40°C temperature difference between its faces. How much heat flows through it per hour? The conductivity $k$ is 0.0025 cal/s · cm · °C.

**226.** What thickness of wood has the same insulating ability as 8 cm of brick? $k = 0.8$ W/m · K for brick and 0.1 W/m · K for wood.

**227.** A boxlike cooler has 5.0-cm-thick walls made of plastic foam. Its total surface area is 1.5 m$^2$. About how much ice melts each hour inside the cooler to hold its temperature at 0°C when the outside temperature is 30°C? Take $k$ for the plastic to be 0.040 W/m · K and $h_v = 80$ cal/g.

**228.** Consider a glass window of area 1 m$^2$ and thickness 0.50 cm. If a temperature difference of 20°C exists between one side and the other, how fast would heat flow through the window? Why is this result *not* applicable to a house window on a day when the temperature difference between inside and outside is 20°C? Take $k = 0.80$ W/m · K.

**229.** A certain double-pane window consists of two glass sheets, each 80 cm × 80 cm × 0.30 cm, separated by a 0.30-cm stagnant air space. The indoor surface temperature is 20°C, while the outdoor surface temperature is 0°C. How much heat passes through the window each second? $k = 2 \times 10^{-3}$ cal/s · cm · °C for glass and about $2 \times 10^{-4}$ cal/s · cm · °C for air.

## Convection

**230.** Distinguish between *natural* and *forced* convection.

**231.** A spherical blackbody of 5-cm radius is maintained at a temperature of 327°C. What is the power radiated?

CHAPTER 14

# Gas Laws and Kinetic Theory

## Gas Laws

**232.** An ideal gas exerts a pressure of 1.52 MPa when its temperature is 298 K (25°C) and its volume is 10 L (10 L).

(A) How many moles of gas are there?
(B) What is the mass density if the gas is molecular hydrogen, $H_2$?
(C) What is the mass density if the gas is oxygen, $O_2$?

**233.** A partially inflated balloon contains 500 $m^3$ of helium at 27°C and 1-atm pressure. What is the volume of the helium at an altitude of 18,000 ft, where the pressure is 0.5 atm and the temperature is –3°C?

**234.** An air bubble released at the bottom of a pond expands to four times its original volume by the time it reaches the surface. If atmospheric pressure is 100 kPa, what is the absolute pressure at the bottom of the pond? Assume constant temperature.

**235.** A pressure gauge indicates the differences between atmospheric pressure and pressure inside the tank. The gauge on a 1.00-$m^3$ oxygen tank reads 30 atm. After some use of the oxygen, the gauge reads 25 atm. How many cubic meters of oxygen at normal atmospheric pressure were used? There is no temperature change during the time of consumption.

**236.** A car tire is filled to a gauge pressure of 24 psi (lb/$in^2$) when the temperature is 20°C. After the car has been running at high speed, the tire temperature rises to 60°C. Find the new gauge pressure within the tire if the tire's volume does not change. (*Note:* Atmospheric pressure is 14.7 psi.)

**237.** In a diesel engine, the cylinder compresses air from approximately standard pressure and temperature to about one-sixteenth the original volume and a pressure of about 50 atm. What is the temperature of the compressed air?

**238.** An ideal gas has been placed in a tank at 40°C. The gauge pressure is initially 608 kPa. One-fourth of the gas is then released from the tank and thermal equilibrium is established. What will be the gauge pressure if the temperature is 315°C? Take standard atmospheric pressure as 101 kPa.

## Kinetic Theory

**239.** What is meant by the *root-mean-square speed* of an ensemble of gas molecules? How is it related to the pressure of an ideal gas?

**240.** Calculate the root-mean-square speed of hydrogen molecules ($H_2$) at 100°C.

**241.** Call the root-mean-square speed of the molecules in an ideal gas $v_0$ at temperature $T_0$ and pressure $p_0$. Find the speed if

(A) the temperature is raised from 20 to 300°C
(B) the pressure is doubled and $T = T_0$
(C) the molecular weight of each of the gas molecules is tripled

**242.** The temperature of outer space has an average value of about 3 K. Find the root-mean-square speed of a proton (a hydrogen nucleus) in space. ($m_p = 1.67 \times 10^{-27}$ kg)

**243.** What is the *mean free path* of a gas molecule?

CHAPTER 15

# The First Law of Thermodynamics

## Basic Thermodynamic Concepts

**244.** Describe the convention for work associated with a thermodynamic system.

**245.** What is meant by the *internal energy* of a system?

**246.** Define *isothermal, isobaric, isovolumic*, and *adiabatic* processes.

**247.** One mole of helium gas, initially at STP ($p_1 = 1$ atm $= 101$ kPa, $T_1 = 0°C$), undergoes an isovolumetric process in which its pressure falls to half its initial value,

(A) What is the work done by the gas?
(B) What is the final temperature of the gas?
(C) The helium gas then expands isobarically to twice its volume; what is the work done by the gas?

## The First Law of Thermodynamics, Internal Energy, *p*-*V* Diagrams, and Cyclical Processes

**248.** What is the change in internal energy of 0.100 mol of nitrogen gas as it is heated from 10 to 30°C at

(A) constant volume?
(B) constant pressure?

**249.** An ideal gas in a cylinder is compressed adiabatically to one-third its original volume. During the process, 45 J of work is done on the gas by the compressing agent.

(A) By how much did the internal energy of the gas change in the process?
(B) How much heat flowed into the gas?

**250.** In each of the following situations, find the change in internal energy of the system.

(A) A system absorbs 2100 J of heat and at the same time does 400 J of work.

(B) A system absorbs 1300 J and at the same time 420 J of work is done on it.

(C) 5000 J of heat is removed from a gas held at constant volume.

**251.** A sample containing 1.00 kmol of the nearly ideal gas helium is put through the cycle of operations shown in Figure 15.1. *BC* is an isothermal, and $p_A$ = 1.00 atm, $V_A$ = 22.4 m$^3$, $p_B$ = 2.00 atm. What are $T_A$, $T_B$, and $V_C$?

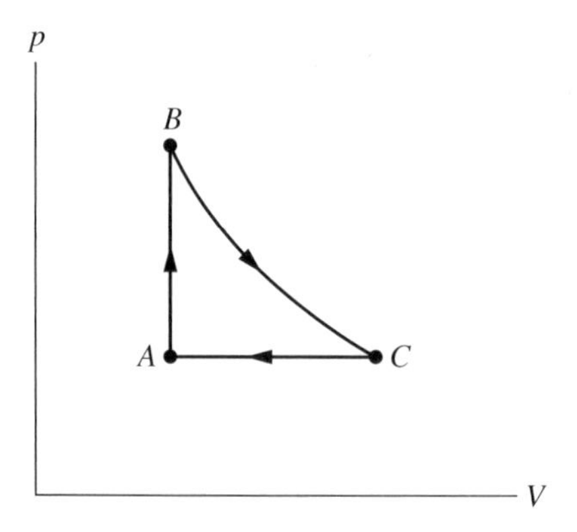


**Figure 15.1**

**252.** The *p*-*V* diagram in Figure 15.2(*a*) represents a reversible cycle of operations performed by an ideal gas in which *MN* is an isothermal and *NK* an adiabatic. Fill in Figure 15.2(*b*) for this cycle, giving the signs for each listed quantity.

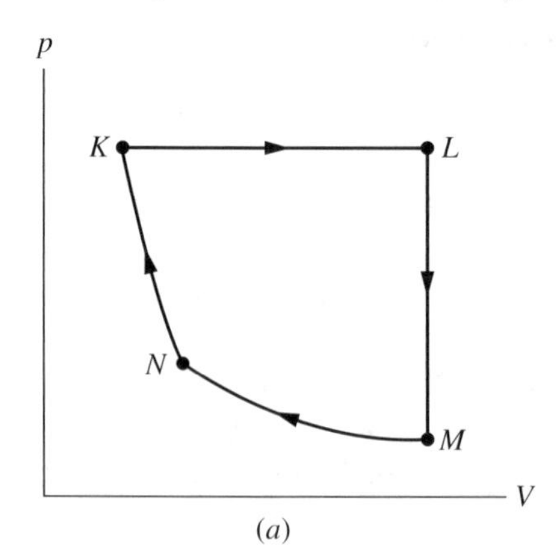


(*a*)

| Path | $\Delta U$ | $Q$ | $W$ | $\Delta T$ |
|---|---|---|---|---|
| *KL* | | | | |
| *LM* | | | | |
| *MN* | | | | |
| *NK* | | | | |

(*b*)

**Figure 15.2**

**253.** The $p$-$V$ diagram shown in Figure 15.3 applies to a gas undergoing a cyclic change in a piston-cylinder arrangement. What is the work done by the gas in

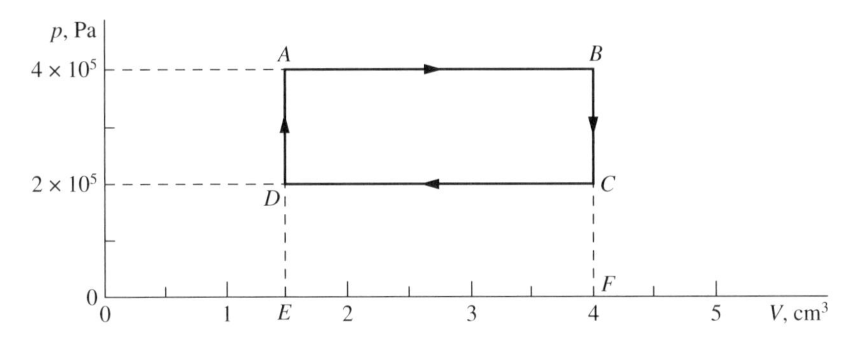


**Figure 15.3**

(A) portion *AB* of the cycle?
(B) portion *BC*?
(C) portion *CD*?
(D) portion *DA*?

**254.** For the thermodynamic cycle shown in Figure 15.3, find

(A) net output work of the gas during the cycle
(B) net heat flow into the gas per cycle

**255.** A cylinder of ideal gas is closed by an 8-kg movable piston (area 60 cm$^2$), as shown in Figure 15.4. Atmospheric pressure is 100 kPa. When the gas is heated from 30 to 100°C, the piston rises 20 cm. The piston is then fastened in place, and the gas is cooled back to 30°C. Calling $Q_1$ the heat added to the gas in the heating process and $|Q_2|$ the heat lost during cooling, find the difference between $Q_1$ and $|Q_2|$.

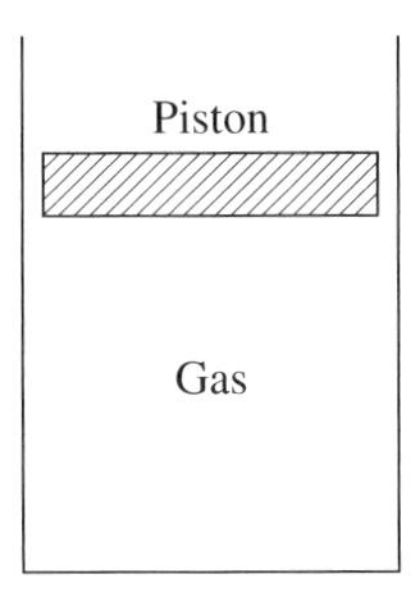


**Figure 15.4**

CHAPTER 16

# The Second Law of Thermodynamics

**256.** What is meant by a *reversible process* in thermodynamics?

**257.** What is a *heat engine,* and what is the *efficiency* of a heat engine?

**258.** A Carnot-type engine is designed to operate between 480 and 300 K. Assuming that the engine actually produces 1.2 kJ of mechanical energy per kilocalorie of heat absorbed, compare the actual efficiency with the theoretical maximum efficiency.

**259.** What is the maximum amount of work that a Carnot engine can perform per kilocalorie of heat input if it absorbs heat at 427°C and exhausts heat at 177°C?

**260.** An ideal Carnot engine takes heat from a source at 317°C, does some external work, and delivers the remaining energy to a heat sink at 117°C. If 500 kcal of heat is taken from the source, how much work is done? How much heat is delivered to the sink?

**261.** An ideal gas is confined to a cylinder by a piston. The piston is slowly pushed in so that the gas temperature remains at 20°C. During the compression, 730 J of work is done on the gas. Find the entropy change of the gas.

**262.** Why isn't the result of Question 261 a violation of the entropy statement of the second law, $\Delta S \geq 0$?

CHAPTER 17

# Wave Motion

## Characteristic Properties

**263.** The sound of a lightning flash is heard 6.0 s after the flash. How far away was the lightning? (The speed of sound is 330 m/s.)

**264.** A radio station broadcasts at 760 kHz. What is the wavelength of the station's radio waves?

**265.** The speed of sound in seawater is 1530 m/s. If a sound of frequency 1800 Hz is produced in seawater, what is its wavelength?

**266.** Sound waves of wavelength $\lambda$ travel from a medium in which their velocity is $v$ into a medium in which their velocity is $4v$. What is the wavelength of the sound in the second medium?

**267.** (A) An ultrasonic transducer used in sonar produces a frequency of 40 kHz. If the speed of the sound wave in seawater is 5050 ft/s, what is its wavelength?
(B) The transducer is made to emit a short burst of sound and is then turned off. The receiver is turned on. The pulse is reflected from a lurking submarine and received 5.0 s after it was first emitted. How far away is the submarine?

**268.** When driven by a 120-Hz vibrator, a string has transverse waves of 31-cm wavelength traveling along it.
(A) What is the speed of the waves on the string?
(B) If the tension in the string is 1.20 N, what is the mass of 50 cm of the string?

**269.** Transverse waves pass along a stretched wire at 1000 ft/s. If the tension on the wire is quadrupled, what will the velocity be?

**270.** The wave shown in Figure 17.1 is being sent out by a 60-Hz vibrator. Find the following for the wave:

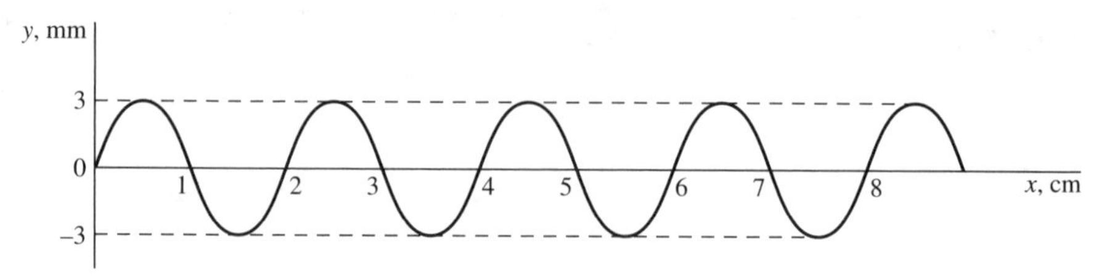


**Figure 17.1**

(A) amplitude
(B) frequency
(C) wavelength
(D) speed
(E) period

**271.** A wave along a string has the following equation ($x$ in meters and $t$ in seconds):

$$y = 0.02 \sin (30t - 4.0x) \text{ m}$$

Find its amplitude, frequency, speed, and wavelength.

**272.** For the wave shown in Figure 17.2, find its amplitude, frequency, and wavelength if its speed is 300 m/s. Write the equation for the wave as it travels out along the $+x$ axis if its position at $t = 0$ is as shown.

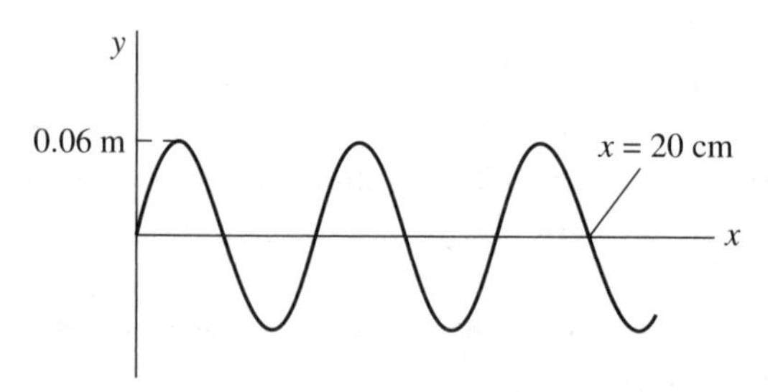


**Figure 17.2**

**273.** A traveling wave on a string has a frequency of 30 Hz and a wavelength of 60 cm. Its amplitude is 2 mm. Find the maximum transverse velocity and maximum transverse acceleration of a point on the string.

**274.** A wire 0.5 m long and with a mass per unit length of 0.0001 kg/m vibrates under a tension of 4 N. Find the fundamental frequency.

**275.** One end of a rubber tube that is 5.0 m long and has a mass-to-length ratio of 0.30 kg/m is fastened to a fixed support; at the other end, a tension of 100 N is applied. If a transverse blow is struck at one end of the tube, how long does it take to reach the other end?

**276.** Standing waves are produced in a rubber tube 12 m long. If the tube vibrates in five segments and the velocity of the wave is 20 m/s, what is

(A) the wavelength of the waves?
(B) the frequency of the waves?

**277.** A string has a length of 0.4 m and a mass of 0.16 g. If the tension in the string is 70 N, what are the three lowest frequencies it produces when plucked?

**278.** The third overtone produced by a vibrating string 2 m long is 1200 Hz. What are the frequencies of the lower overtones and of the fundamental? What is the velocity of propagation?

**279.** A 160-cm long string has two adjacent resonances at frequencies of 85 and 102 Hz.

(A) What is the fundamental frequency of the string?
(B) What is the length of a segment at the 85-Hz resonance?
(C) What is the speed of the waves on the string?

**280.** An 800 g mass is suspended vertically by a 200 cm length of string. The string is found to resonate in three segments to a frequency of 480 Hz. What is the mass per unit length of string?

**281.** An organ pipe 0.30 m long is open at both ends. What are the frequencies of the fundamental and of the first two overtones?

**282.** A 0.77 m long organ pipe is closed at one end. What is its fundamental frequency, and what are the frequencies of the first two overtones?

**283.** Sound waves of frequency 320 Hz are sent into the top of a vertical tube containing water at a level that can be adjusted. If standing waves are produced at two successive water levels—20 cm and 73 cm—what is the speed of the sound waves in the air of the tube?

**284.** A 40-cm long brass rod is dropped one end first onto a hard floor but is caught before it topples over. With an oscilloscope, it is determined that the impact produces a 3-kHz tone. What is the speed of sound in brass?

CHAPTER 18

# Sound

## Sound Velocity; Beats; Doppler Shift

**285.** A 1000-Hz sound wave in air strikes the surface of a lake and penetrates the water. What are the frequency and wavelength of the wave in water? The speed of sound in water is 1500 m/s.

**286.** An underwater sonar source operating at a frequency of 60 kHz directs its beam toward the surface. What is the wavelength of the beam in the air above? What frequency sound due to the sonar source does a bird flying above the water hear? The speed of sound in air is 330 m/s.

**287.** Two closed organ pipes sounded simultaneously give five beats per second between the fundamentals. If the shorter pipe is 1.1 m long, find the length $L$ of the longer pipe. The speed of sound in air is 340 m/s.

**288.** Two open organ pipes, one 2.5 ft and one 2.4 ft in length, are sounded simultaneously. How many beats per second will be produced between the fundamental tones if the speed of the sound is 1100 ft/s?

**289.** Four beats per second are heard when two tuning forks are sounded simultaneously. After attaching a small piece of tape to one prong of the second tuning fork, the two tuning forks are sounded again and two beats per second are heard. If the first fork has a frequency of 180 Hz, what must the original frequency of the second fork have been?

**290.** A train is moving toward an observer with a speed of 100 ft/s (68 mi/h). The whistle of the locomotive has a frequency of 400 Hz, and the speed of the sound is 1100 ft/s. Find the frequency heard by the observer.

**291.** A car is traveling at a speed of 90 ft/s (61 mi/h) along a road paralleling a railroad track. Practically straight in front of the car is a locomotive waiting on a siding. If the speed of the sound is 1080 ft/s and the driver of the car hears a frequency of 400 Hz when the locomotive whistle is blown, what is the actual frequency emitted by the whistle?

**292.** A hawk is flying directly away from a birdwatcher and directly toward a distant cliff at a speed of 15 m/s. The hawk produces a shrill cry whose frequency is 800 Hz.

(A) What is the frequency in the sound that the birdwatcher hears directly from the bird?

(B) What is the frequency that the birdwatcher hears in the echo that is reflected from the cliff?

CHAPTER 19

# Coulomb's Law and Electric Fields

## Coulomb's Law of Electrostatic Force

**293.** A test charge $Q = +2\ \mu C$ is placed halfway between a charge $Q_1 = +6\ \mu C$ and a charge $Q_2 = +4\ \mu C$, which are 10 cm apart. Find the force on the test charge and its direction.

**294.** Three +20-$\mu$C charges are placed along a straight line, successive charges being 2 m apart, as shown in Figure 19.1. Calculate the force on the charge on the right end.

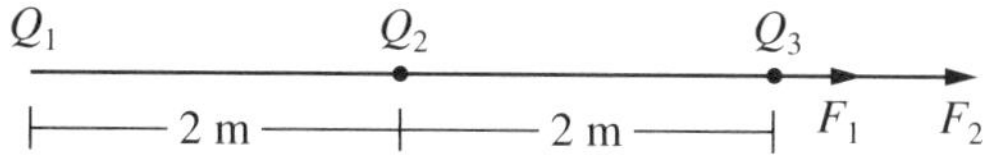

$Q_1$ $Q_2$ $Q_3$

2 m 2 m $F_1$ $F_2$

**Figure 19.1**

**295.** Four equal point changes, +3 $\mu$C, are placed at the four corners of a square that is 40 cm on a side. Find the force on any one of the charges.

**296.** Four equal-magnitude point charges (3 $\mu$C) are placed at the corners of a square that is 40 cm on a side. Two, diagonally opposite each other, are positive, and the other two are negative. Find the force on either negative charge.

**297.** Charges of +2, +3, and −8 $\mu$C are placed at the vertices of an equilateral triangle of side 10 cm. Calculate the magnitude of the force acting on the −8 $\mu$C charge due to the other two charges.

**298.** One charge (+5 $\mu$C) is placed at $x = 0$, and a second charge (+7 $\mu$C) at $x = 100$ cm. Where can a third be placed and experience zero net force due to the other two?

**299.** Two identical tiny metal balls carry changes of +3 nC and −12 nC. They are 3 cm apart.

(A) Compute the force of attraction.

(B) The balls are now touched together and then separated to 3 cm. Describe the forces on them now.

**300.** The two balls shown in Figure 19.2 have identical masses of 0.20 g each. When suspended from 50-cm-long strings, they make an angle of 37° to the vertical. If the charges on each are the same, how large is each charge?

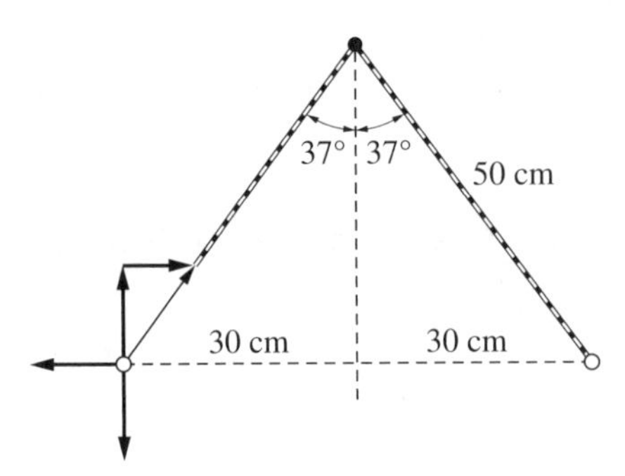


**Figure 19.2**

**301.** In the Bohr model, taking the radius of the hydrogen atom as $r = 5.29 \times 10^{-11}$ m, find the strength of the force on the electron, its centripetal acceleration, and its orbital speed.

## The Electric Field, Continuous Charge Distributions, and Motion of Charged Particles in an Electric Field

**302.** A charge, +6 $\mu$C, experiences a force of 2 mN in the +$x$ direction at a certain point in space.

(A) What was the electric field there before the charge was placed?

(B) Describe the force a −2 $\mu$C charge would experience if it were used in place of the +6 $\mu$C.

**303.** Find the electric field at point $P$ in Figure 19.3 due to the charges shown. Draw a diagram without electric fields; add the fields in the solution.

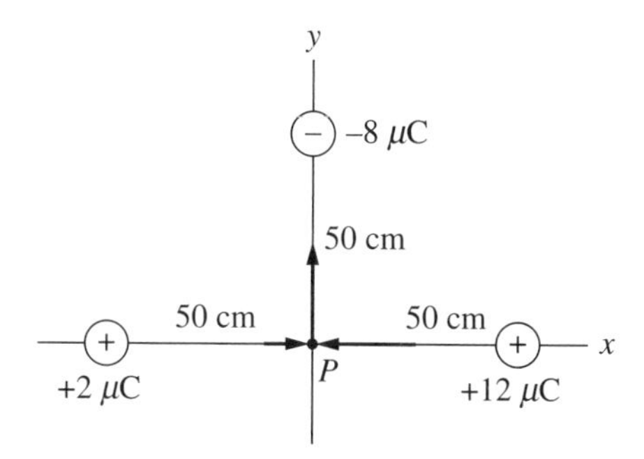


**Figure 19.3**

**304.** Four equal-magnitude (4-$\mu$C) charges are placed at the four corners of a square that is 20 cm on each side. Find the electric-field intensity at the center of the square, if the charges are all positive.

**305.** Rework Question 304 if the charges alternate in sign around the perimeter of the square.

**306.** Three charges are placed at three corners of a square, as shown in Figure 19.4. Find the electric-field strength at point $A$ (magnitude and direction).

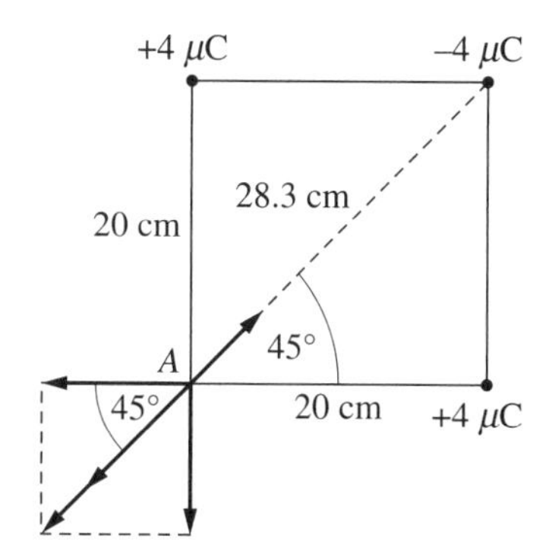


**Figure 19.4**

**307.** A tiny 0.60-g ball carries a charge of magnitude 8 $\mu$C. It is suspended by a thread in a downward electric field of intensity 300 N/C. What is the tension in the thread if the charge on the ball is

(A) positive?
(B) negative?

**308.** The tiny ball at the end of the thread has a mass of 0.60 g and is in a horizontal electric field of intensity 700 N/C. It is in equilibrium in the position shown in Figure 19.5. What are the magnitude and sign of the charge on the ball?

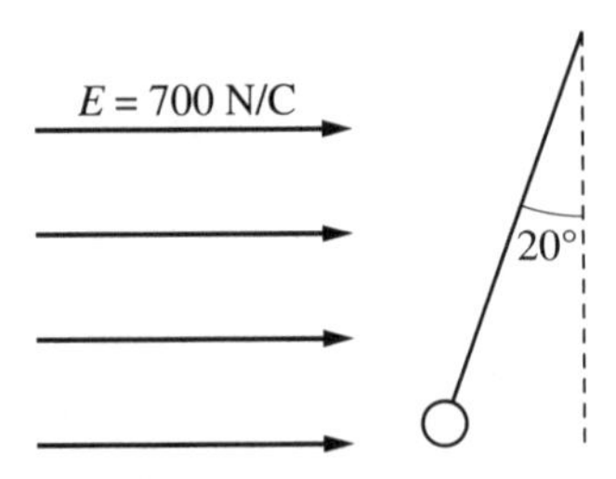


**Figure 19.5**

**309.** An electron ($q = 1.6 \times 10^{-19}$ C, $m = 9.1 \times 10^{-31}$ kg) is projected out along the $+x$ axis with an initial speed of $3 \times 10^6$ m/s. It goes 45 cm and stops due to a uniform electric field in the region. Find the magnitude and direction of the field.

**310.** An electron is shot at $1 \times 10^6$ m/s between two parallel charged plates, as shown in Figure 19.6. If $E$ between the plates is 1 kN/C, where will the electron strike the upper plate? Assume vacuum conditions.

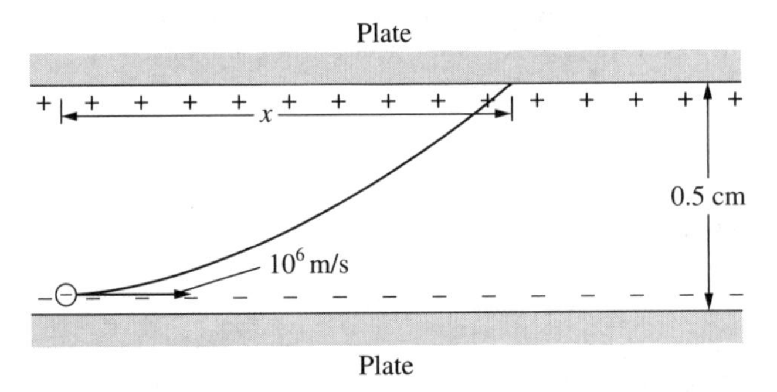


**Figure 19.6**

**311.** An oil drop carries six electronic charges, has a mass of $1.6 \times 10^{-12}$ g, and falls with a terminal velocity in air. What magnitude of vertical electric field is required to make the drop move upward with the same speed as it was formerly moving downward?

CHAPTER 20

# Electric Potential and Capacitance

## Potential Due to Point Charges or Charge Distributions

**312.** A point charge, $q_1 = +2\ \mu C$,is placed at the origin of coordinates. A second, $q_2 = -3\ \mu C$, is placed on the $x$ axis at $x = +100$ cm. At what point (or points) on the $x$ axis will the absolute potential be zero?

**313.** Three equal charges of +6 nC are located at the corners of an equilateral triangle whose sides are 12 cm long (Figure 20.1). Find the potential at the center of the base of the triangle.

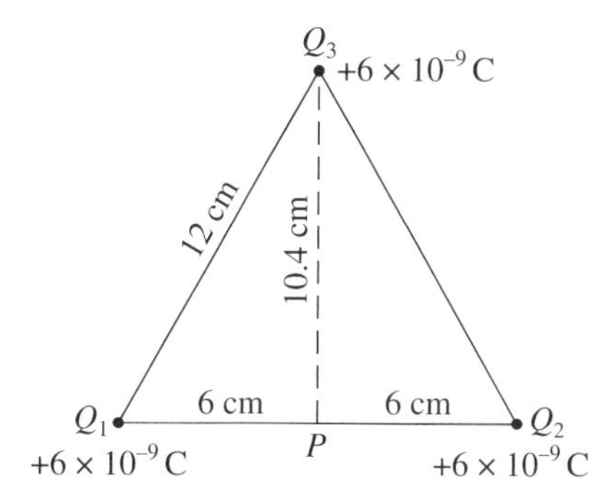


**Figure 20.1**

**314.** A potential difference of 150 V is applied to two parallel metal plates. If an electric field of 5000 V/m is produced between the plates, how far apart are the plates?

**315.** The charge on an electron is $1.6 \times 10^{-19}$ C in magnitude. An oil drop has a weight of $3.2 \times 10^{-13}$ N. With an electric field of $5 \times 10^5$ V/m between the plates of Millikan's oil-drop apparatus, this drop is observed to be essentially balanced. What is the charge on the drop in electronic charge units?

**316.** In the Millikan experiment, an oil drop carries four electronic charges and has a mass of $1.8 \times 10^{-12}$ g. It is held almost at rest between two horizontal charged plates 1.8 cm apart. What voltage must there be between the two charged plates?

**317.** Two large parallel metal plates (3.00 mm apart) are charged to a potential difference of 12 V.

(A) What is the field between them?
(B) They are now disconnected from the battery and pulled apart to 5.00 mm. What is the new electric field between them and what is now the potential difference?

**318.** A pair of horizontal metal plates are separated by a vertical distance of 10 cm, and the voltage difference between them is 28 V (Figure 20.2). A small ball of 0.60-g mass hangs by a thread from the upper plate. What is the tension in the thread if the ball carries a charge of 20 m $\mu$C? Two answers are possible. Find both.

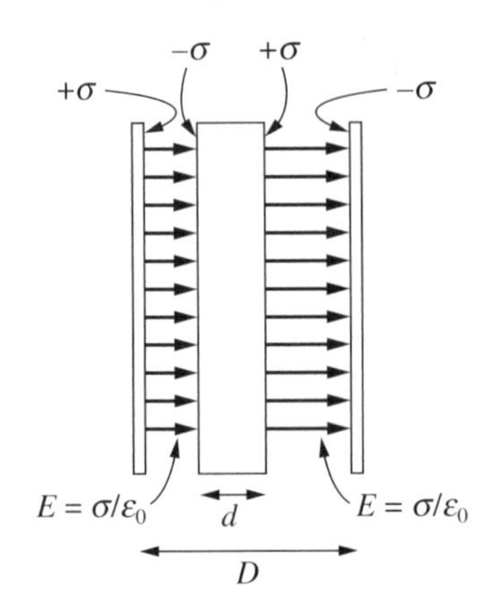


**Figure 20.2**

**319.** An electron gun shoots electrons ($q = -e$, $m = 9.1 \times 10^{-31}$ kg) at a metal plate that is 4 mm away in vacuum. The plate is 5.0 V lower in potential than the gun. How fast must the electrons be moving as they leave the gun if they are to reach the plate?

**320.** The electron beam in a television tube consists of electrons accelerated from rest through a potential difference of about 20 kV. How large an energy do the electrons have? What is their speed? Ignore relativistic effects for this approximate calculation. ($m_e = 1.9 \times 10^{-31}$ kg)

**321.** Two metal plates are attached to the two terminals of a 1.50-V battery in a vacuum. An electron ($q = -e$, $m = 9.1 \times 10^{-31}$) is released at the negative plate and falls freely to the positive plate. How fast is it going just before it strikes the plate?

**322.** A proton is released from a point $P$, which is $10^{-14}$ m from a heavy nucleus that has a charge of $80e$. (This would be the nucleus of a mercury atom.) How large will the kinetic energy of the proton be when it gets far away from the nucleus? What will be its speed?

**323.** The potential difference between the two plates in Figure 20.3 is 100 V. If the system is in vacuum, what will be the speed of a proton released from plate $B$ just before it hits plate $A$?

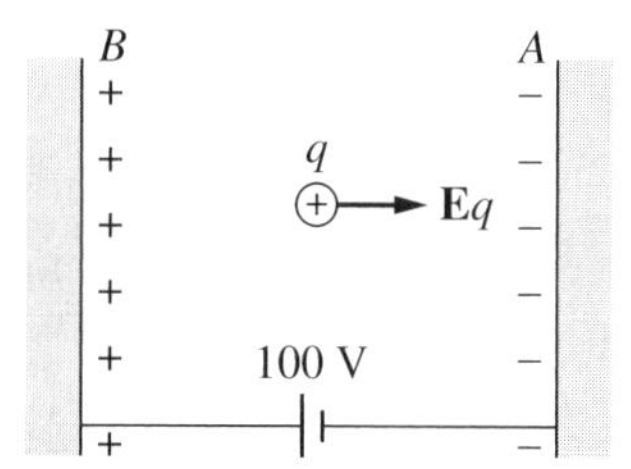


**Figure 20.3**

**324.** A certain parallel-plate capacitor consists of two plates, each with area 200 cm$^2$, separated by a 0.4-cm air gap.

(A) Compute its capacitance.
(B) If the capacitor is connected across a 500-V source, what are the charge on it, the energy stored in it, and the value of $E$ between the plates?
(C) If a liquid with $K = 2.60$ is poured between the plates so as to fill the air gap, how much additional charge will flow onto the capacitor from the 500-V source?

**325.** A 5-$\mu$F capacitor with air between the metal plates is connected to a 30-V battery. The battery is then removed, leaving the capacitor charged.

(A) Calculate the charge on the capacitor.
(B) The air between the plates is replaced by oil with $K = 2.1$. Find the new value of the capacitance and the new potential difference between the plates.

**326.** The series combination of two capacitors shown in Figure 20.4 is connected across 1000 V. Compute

(A) the equivalent capacitance $C_{eq}$ of the combination
(B) the magnitude of the charge on each capacitor
(C) the potential differences across each capacitor
(D) the energy stored on each capacitor

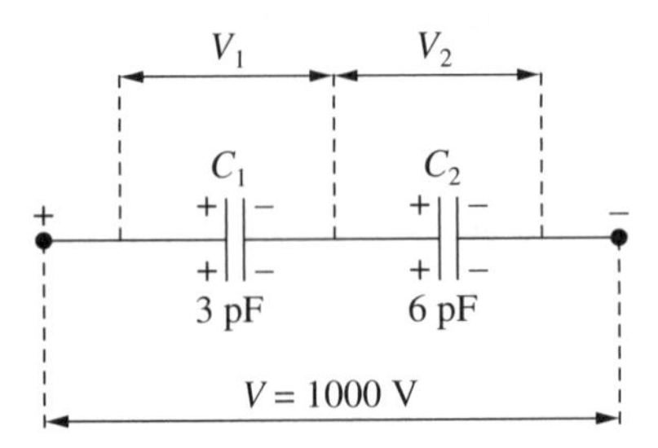


**Figure 20.4**

**327.** Three capacitors (2, 3, and 4 $\mu$F) are connected in series with a 6-V battery. When the current stops, what is the charge on the 3-$\mu$F capacitor? What is the potential difference between the two ends of the 4-$\mu$F capacitor?

**328.** Three capacitors are connected as shown in Figure 20.5. A 12-V potential difference is applied to the terminals.

(A) What is the equivalent capacitance of the arrangement?
(B) Determine the charge on each capacitor.
(C) Determine the voltage on each capacitor.

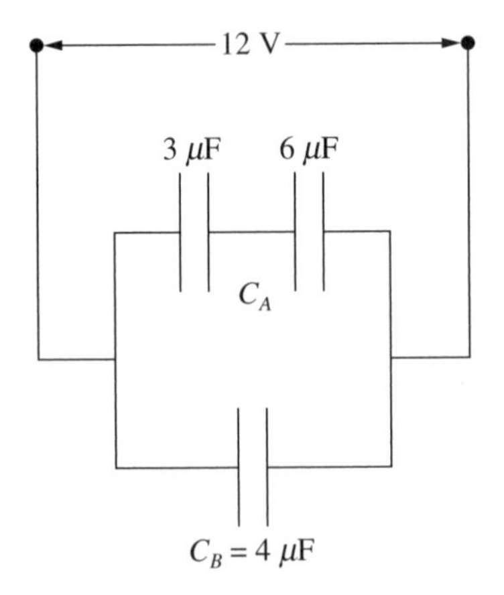


**Figure 20.5**

**329.** (A) In the circuit of Figure 20.6, find the equivalent capacitance.
(B) Determine the potential difference across each capacitor.

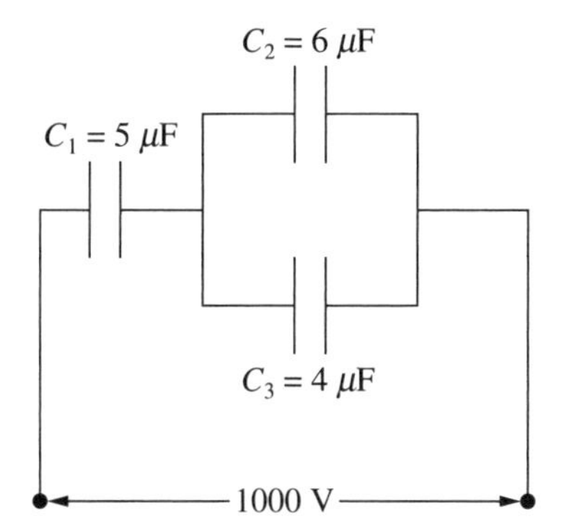


**Figure 20.6**

**330.** Two capacitors in parallel, 2 and 4-$\mu$F, are connected, as a unit, in series with a 3-$\mu$F capacitor (see Figure 20.7). The combination is connected across a 12-V battery. Find the equivalent capacitance of the combination and the potential difference across the 2-$\mu$F capacitor.

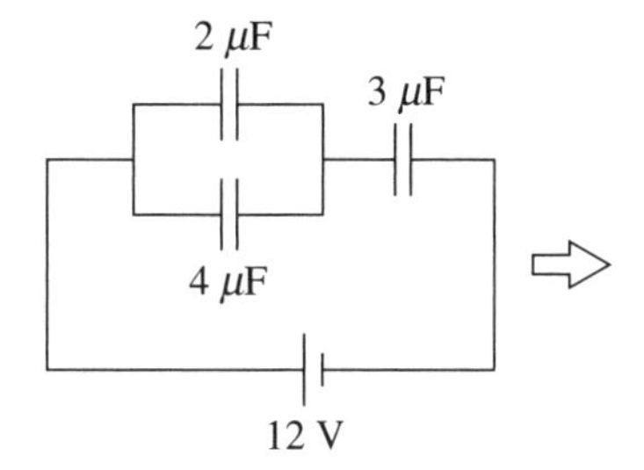


**Figure 20.7**

**331.** Two capacitors, 3 $\mu$F and 4 $\mu$F, are individually charged across a 6-V battery. After being disconnected from the battery, they are connected together with the negative plate of one attached to the positive plate of the other. What is the final charge on each capacitor?

**332.** Two capacitors, $C_1 = 4\ \mu F$ and $C_2 = 6\ \mu F$, are originally connected to a battery $V = 12$ V, as shown in Figure 20.8, and then disconnected and reconnected as shown. What is the final charge on each capacitor?

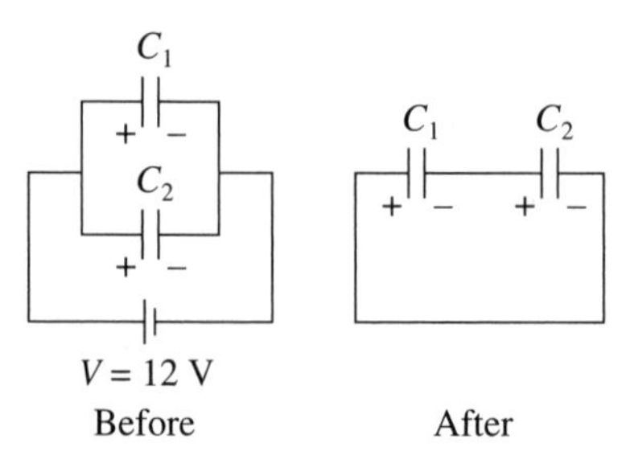


**Figure 20.8**

CHAPTER 21

# Simple Electric Circuits

## Ohm's Law, Current, Resistance

**333.** Determine the potential difference between the ends of a wire of resistance 5 Ω if 720 C passes through it per minute.

**334.** How long a piece of aluminum wire 1 mm in diameter is needed to give a resistance of 4 Ω at room temperature?

**335.** Three resistances of 12, 16, and 20 ω are connected in parallel. What resistance must be connected in series with this combination to give a total resistance of 25 ω?

**336.** Find all the resistances that can be realized with a 6-, a 9-, and a 15-ω resistor in various combinations. Not every combination need use all three resistors.

**337.** Arrange an 8-, a 12-, and a 16-ω resistor in a combination that has a total resistance of 8.9 ω.

**338.** Suppose that the emf of a battery in Figure 21.1 is 45 V and the resistor $R_1 = 300\ \Omega$.

(A) What must the resistor $R_2$ be in order that the current $I$ be 0.45 A?

(B) What are the currents $I_1$ and $I_2$?

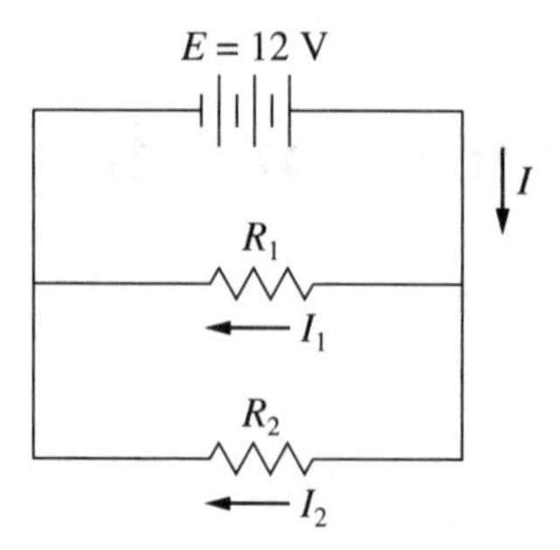


**Figure 21.1**

**339.** There are three resistors in Figure 21.1 are $R_1 = 25\ \Omega$, $R_2 = 50\ \Omega$, and $R_3 = 100\ \Omega$

(A) What is the total resistance of the circuit?

(B) What are the currents $I_1$, $I_2$, and $I_3$ for a 12-V battery?

**340.** The three resistors in Figure 21.2 are $R_1 = 80\ \Omega$, $R_2 = 25\ \Omega$, and $R_3 = 15\ \Omega$

(A) What is the total resistance of the circuit?

(B) What are the currents $I$ and $I_2$, and the voltage across the battery, if $I_1 = 0.3$ A?

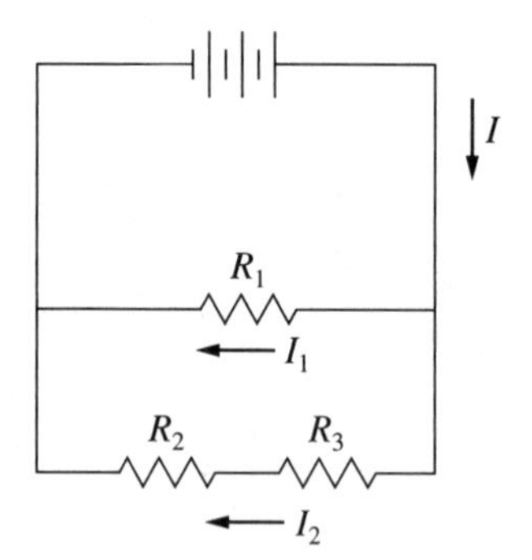


**Figure 21.2**

**341.** Figure 21.3 shows the three resistors $R_1 = 5\ \Omega$, $R_2 = 15\ \Omega$, and $R_3 = 25\ \Omega$ in four different circuits. For each circuit find the currents $I_1$, $I_2$, and $I_3$ in each resistor, and the current $I$ in the battery.

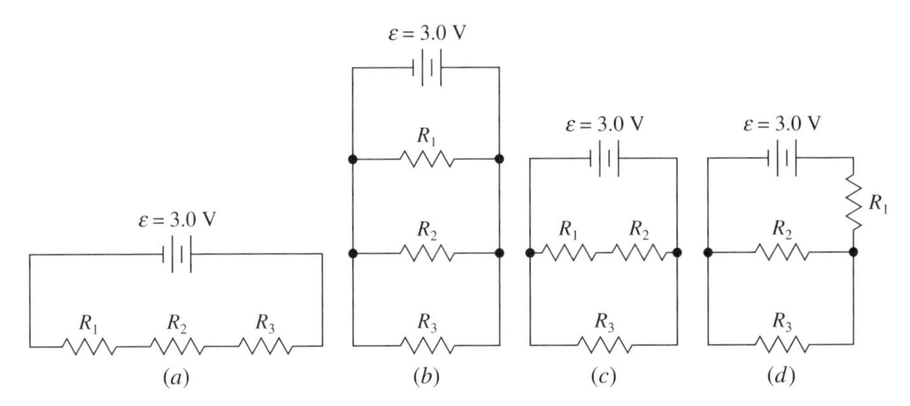


**Figure 21.3**

**342.** A battery usually has a small internal resistance of its own. This is indicated by the resistor $r$ in Figure 21.4. If the emf $\mathcal{E}$ of the battery is 3.0 V, $r = 0.5\ \Omega$, and $R = 5\ \Omega$, what is the potential difference between the terminals $a$ and $b$ of the battery?

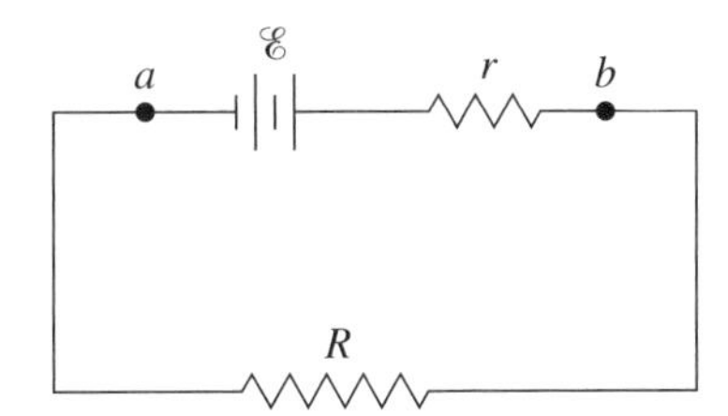


**Figure 21.4**

**343.** The circuit in Figure 21.4 has a current of 0.5 A when $R$ is 10 Ω and a current of 0.27 A when $R$ is 20 Ω. Find

(A) the internal resistance $r$
(B) the $\mathcal{E}$ of the battery

**344.** Compute the internal resistance of an electric generator that has an emf of 120 V and a terminal voltage of 110 V when supplying 20 A.

**345.** A battery charger supplies a current of 10 A to charge a storage battery that has an open-circuit voltage of 5.6 V. If the voltmeter connected across the charger reads 6.8 V, what is the internal resistance of the battery at this time?

**346.** An automobile battery has an emf of 6 V and an internal resistance of 0.01 Ω. When the starter draws 200 A from the battery, what is the terminal voltage of the battery?

**347.** A current of 0.50 A flows through a 200-Ω resistor. How much power is dissipated by the resistor?

**348.** A bulb rated 90 W at 120 V is operated from a 120-V power source. Find the current flowing through it and its resistance.

**349.** A 4 ohm light bulb is connected to a 12 V battery.
(A) What is the power dissipated in the bulb?
(B) What is the power dissipated in a 2 ohm light bulb connected to the same battery?
(C) Which bulb is brighter?

**350.** A 500-W heater is used to heat 250 mL of water from 20 to 100°C. What is the minimum time in which this can be done?

**351.** It is desired to heat a cup of coffee (200 mL) by use of an immersion heater from 20 to 90°C in 0.5 min. How much current would the heater draw from 120 V?

## More Complex Circuits, Kirchhoff's Circuit Rules, and Circuits with Capacitors

**352.** State Kirchhoff's laws.

**353.** In Figure 21.5, find $I_1$, $I_2$, and $I_3$ if switch $k$ is open.

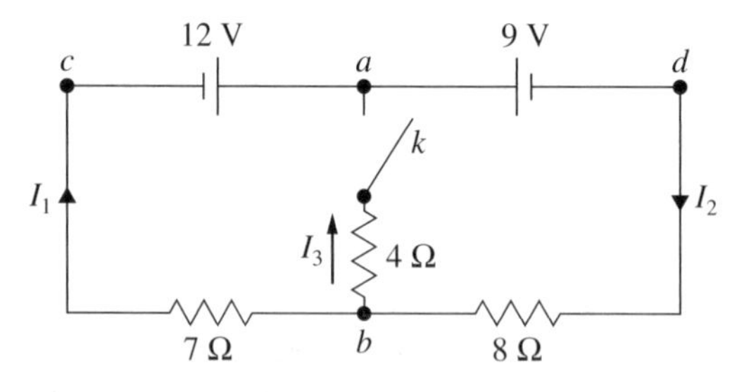


**Figure 21.5**

**354.** In Figure 21.5, find $I_1$, $I_2$, $I_3$ if switch $k$ is closed.

CHAPTER 22

# The Magnetic Field

## Force on a Moving Charge

**355.** The particle shown in Figure 22.1 is positively charged. What is the direction of the force on it due to the magnetic field? Give its magnitude in terms of $B$, $q$, $\theta$ and $v$.

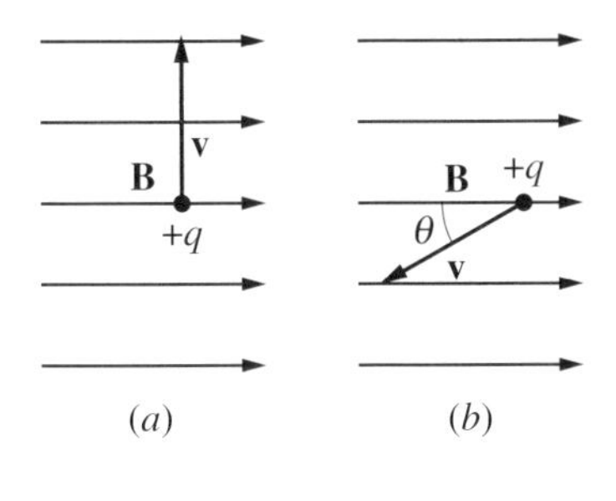


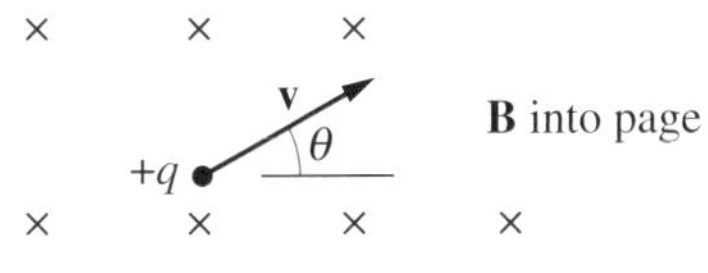


(c)

**Figure 22.1**

**356.** A beam of protons (positively charged) is moving horizontally toward you. As it approaches, it passes through a magnetic field directed downward. This magnetic field deflects the beam to your ______________.

**357.** An electron is accelerated from rest through a potential difference of 3750 V. It enters a region where $B = 4$ mT perpendicular to its velocity. Calculate the radius of the path it will follow.

**358.** Alpha particles ($m = 6.68 \times 10^{-27}$ kg, $q = +2e$), accelerated through a potential difference $V$ to 2 keV, enter a magnetic field $B = 0.2$ T perpendicular to their direction of motion. Calculate the radius of their path.

**359.** A particle with charge $+q$ and mass $m$ is shot with kinetic energy $K$ into the region between two plates separated by a distance $d$, as shown in Figure 22.2. If the magnetic field between the plates is $B$ and as shown, how large must $B$ be if the particle is to miss collision with the opposite plate?

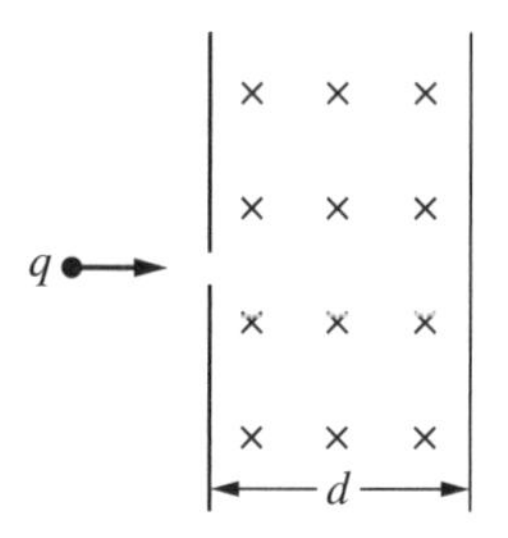


**Figure 22.2**

**360.** In Figure 22.3, a proton ($q = +e$, $m = 1.67 \times 10^{-27}$ kg) is shot with speed $8 \times 10^6$ m/s at an angle of 30° to an $x$-directed field $B = 0.15$ T. Describe the path followed by the proton.

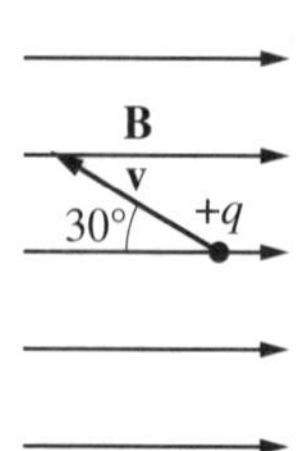


**Figure 22.3**

**361.** In Nebraska, the horizontal component of the earth's field is $2 \times 10^{-5}$ T. If a vertical wire carries a current of 30 A upward there, what is the magnitude and direction of the force on 1 m of the wire?

**362.** As shown in Figure 22.4, a conducting bar of mass $M$ is suspended by two springs. Assume that a magnetic field **B** is directed out of the page. Each spring has a spring constant $k$. Describe the bar's displacement when a current $I$ is sent through it in the direction shown.

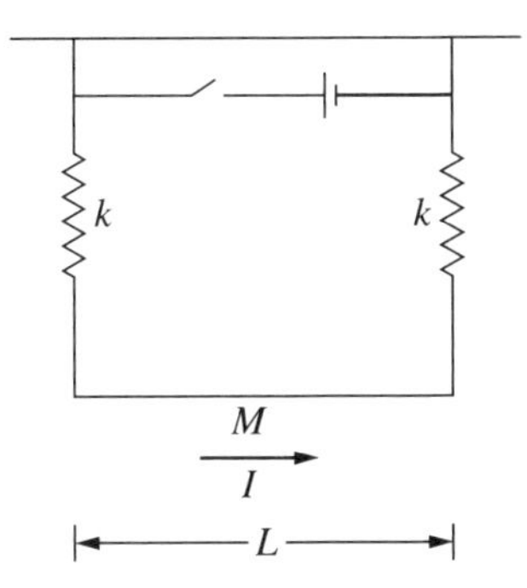


**Figure 22.4**

**363.** Figure 22.5 shows two long parallel wires separated by a distance of 180 mm. There is a current of 8 A in wire 1 and a current of 12 A in wire 2.

(A) Find the total magnetic field at point *A*, which is on the line joining the wires and 30 mm from wire 1 and 150 mm from wire 2.

(B) At what point on the line joining the wires is the magnetic field zero?

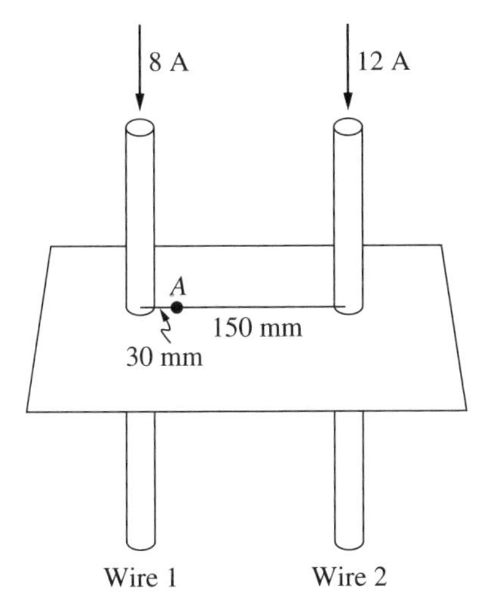


**Figure 22.5**

**364.** A horizontal wire is lined up in the north-south direction. A compass needle is placed above the wire. When the current is turned on, the North pole of the compass is deflected toward the west. In which direction does the current in the wire flow?

**365.** Two long and fixed parallel wires, *A* and *B*, are 10 cm apart in air and carry currents of 40 and 20 A, respectively, in opposite directions. Determine the resultant magnetic field

(A) on a line midway between the wires and parallel to them
(B) on a line 8 cm from wire *A* and 18 cm from wire *B*

**366.** Consider the three long, straight, parallel wires. Find the force experienced by a 25-cm length of wire *C* (see Figure 22.6).

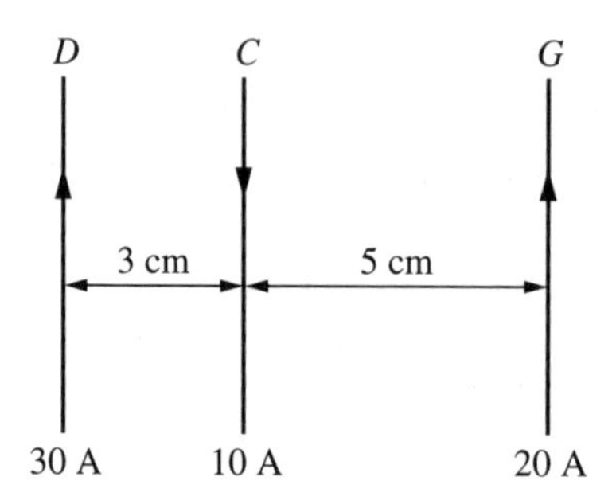


**Figure 22.6**

**367.** For the situation shown in Figure 22.7,

(A) find the force experienced by side *MN* of the rectangular loop.
(B) find the torque on the loop.

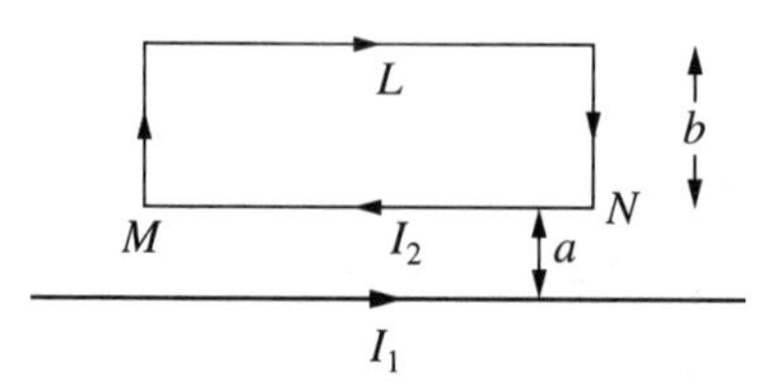


**Figure 22.7**

CHAPTER 23

# Induced emf; Generators and Motors

**368.** In Figure 23.1 there is a $+x$-directed magnetic field of 0.2 T. Find the magnetic flux through each face of the box.

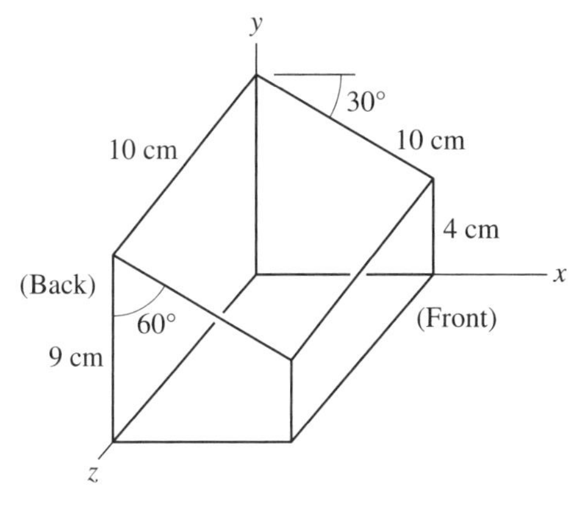


**Figure 23.1**

**369.** The quarter-circle loop shown in Figure 23.2 has an area of 15 cm$^2$. A magnetic field, with $B = 0.16$ T, exists in the $+x$ direction. Find the flux through the loop in each orientation shown.

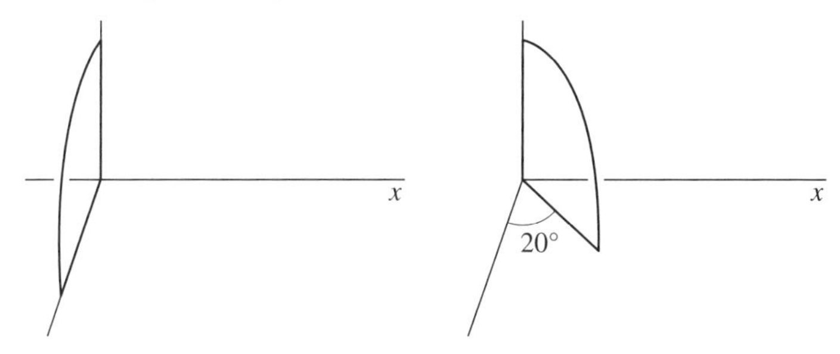


**Figure 23.2** (*Cont.*)

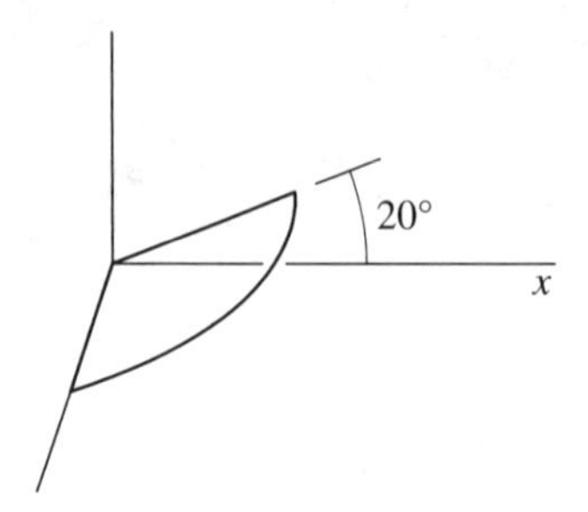


**Figure 23.2** (*Cont.*)

**370.** The perpendicular component of the external magnetic field through a 10-turn coil of radius 50 mm increases from 0 to 18 T in 3 s, as shown in Figure 23.3. If the resistance of the coil is 2 Ω, what is the magnitude of the induced current? What is the direction of the current?

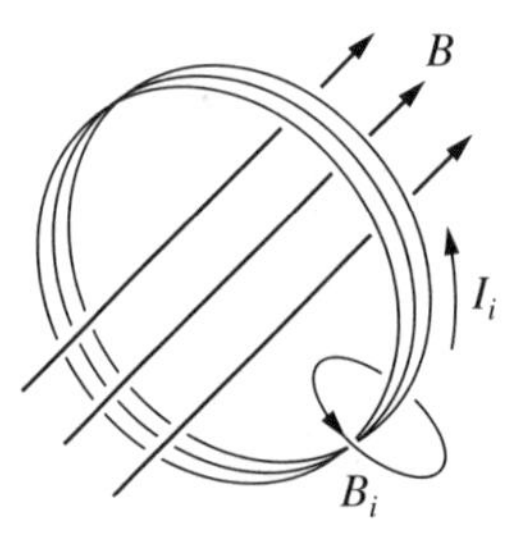


**Figure 23.3**

**371.** A 50-loop circular coil has a radius of 30 mm. It is oriented so that the field lines of a magnetic field are parallel to a normal to the area of the coil. Suppose that the magnetic field is varied so that $B$ increases from 0.10 to 0.35 T in a time of 2 ms. Find the average induced emf in the coil.

**372.** In Figure 23.4 the rectangular loop of wire is being pulled to the right, away from the long, straight wire through which a steady current $i$ flows upward. Does the current induced in the loop flow in the clockwise sense or in the counterclockwise sense?

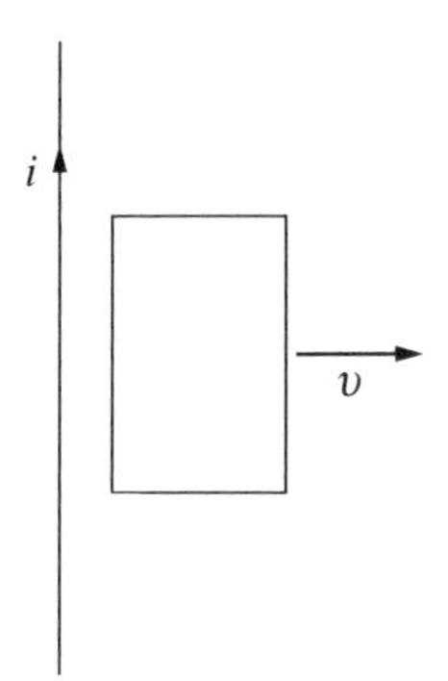


**Figure 23.4**

**373.** An aluminum ring is placed around the projecting core of a powerful electromagnet. See Figure 23.5. When the circuit is closed, the ring jumps up to a surprising height. Explain.

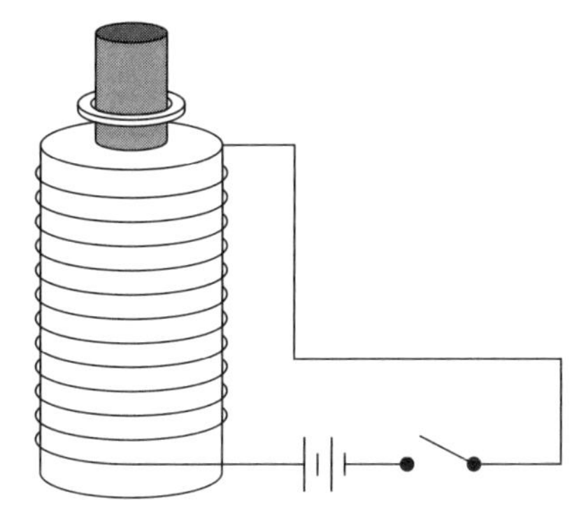

**Figure 23.5**

**374.** A pendulum consists of a pivoted rod at the lower end of which there is a metal ring. The pendulum swings in the plane of the ring. The ring is raised and released. At the bottom of its swing, the ring enters a magnetic field normal to the plane of the ring, in the gap of a strong horseshoe magnet. On entering this region, it soon comes to a stop. Why?

**375.** For the situation of Question 374, if a small piece of the ring is cut out and the experiment repeated, the pendulum keeps swinging for some time. Explain.

**376.** The magnet in Figure 23.6 induces an emf in the coils as the magnet moves toward the right or the left. Find the directions of the induced currents through the resistors when the magnet is moving

(A) toward the right
(B) toward the left.

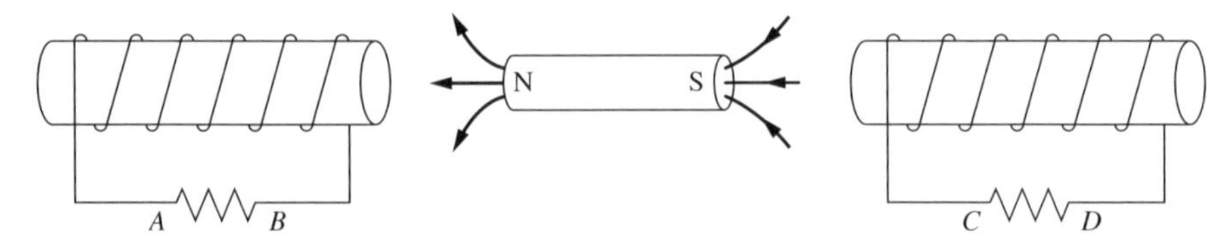


**Figure 23.6**

**377.** The magnet in Figure 23.7 rotates as shown on a pivot through its center. At the instant shown, what are the directions of the induced currents?

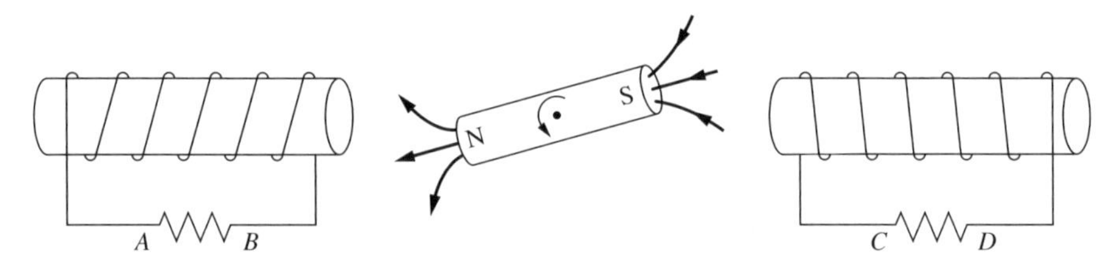


**Figure 23.7**

**378.** A square loop of wire 75 mm on a side lies with its plane perpendicular to a uniform magnetic field of 0.8 T.

(A) Find the magnetic flux through the loop.
(B) If the coil is rotated through 90° in 0.015 s in such a way that there is no flux through the loop at the end, find the average emf induced during the rotation.

**379.** Calculate the induced emf in a 150-cm$^2$ circular coil having 100 turns when the field strength $B$ passing through the coil changes from 0.0 to 0.001 T in 0.1 s at a constant rate.

**380.** A coil of 275 turns with an area of 0.024 m$^2$ is placed with its plane perpendicular to the earth's field and is rotated in 0.025 s through a quarter turn, so that its plane is parallel to the earth's field. What is the average emf induced if the earth's field has an intensity of 80 $\mu$T? What was the original flux through each turn?

## Motional emf; Induced Currents and Forces

**381.** A conductor of length 0.19 m is passing perpendicular to a magnetic field of intensity 0.003 $Wb/m^2$ at a velocity of 11.5 m/s, as shown in Figure 23.8. Calculate the induced emf.

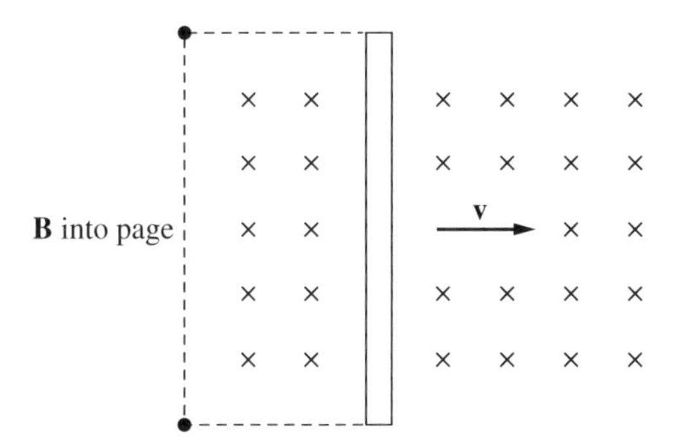


**Figure 23.8**

**382.** As shown in Figure 23.9, a metal rod makes contact with a partial circuit and completes the circuit. The circuit area is perpendicular to a magnetic field with $B = 0.15$ T. If the resistance of the total circuit is 3 Ω, how large a force is needed to move the rod as indicated with a constant speed of 2 m/s?

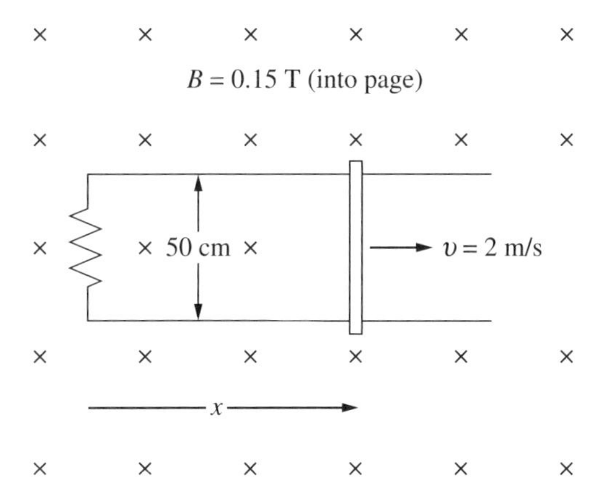


**Figure 23.9**

## Electric Generators and Motors

**383.** A hand-operated generator is easy to turn when it is not connected to any electric device. However, it becomes quite difficult to turn when it is connected, particularly if the device has a low resistance. Explain.

**384.** Determine the separate effects on the induced emf of a generator if

(A) the flux per pole is doubled
(B) the speed of the armature is doubled

**385.** Figure 23.10 represents a primitive motor. A metal wire slides on a horseshoe-shaped metal loop of width 0.25 m. These have negligible resistance, but there is a 1.0-Ω resistor in the circuit as well as a 6.0-V battery. There is a uniform magnetic field directed into the plane of the page of magnitude 0.50 T. The slide wire is pushed to the right by the magnetic force. A force of 0.25 N to the left is required to keep it moving with constant speed to the right.

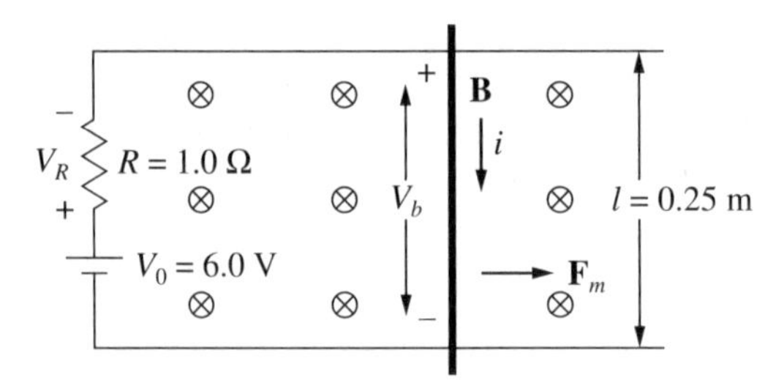


**Figure 23.10**

(A) What is the current $i$ in the circuit?
(B) What is the voltage drop across the 1.0-Ω resistor?
(C) What is the back emf generated by the moving wire?

**386.** Refer to Question 385.

(A) With what constant speed is the wire moving?
(B) What mechanical power does the motor produce?
(C) What is the efficiency of the motor?

CHAPTER 24

# Electric Circuits

**387.** Describe the charging process for an *R-C* circuit (see Figure 24.1) and the time constant of the circuit. Repeat for the discharging process when the battery is removed.

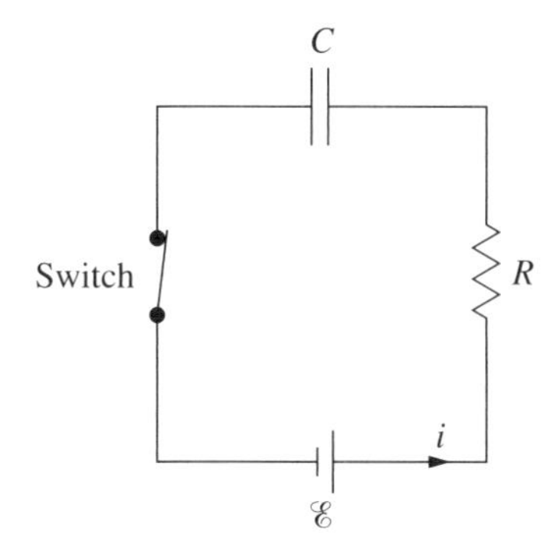


**Figure 24.1**

**388.** A series circuit consists of a 12-V battery, a switch, a 1-MΩ resistor, and a 2-μF capacitor, initially uncharged. If the switch is now closed, find the

(A) initial current in the circuit
(B) time taken for the current to drop to 0.37 of its initial value
(C) charge on the capacitor at the time found in (B)
(D) charge on the capacitor after a long time

**389.** A 5-$\mu$F capacitor is charged to a potential of 20 kV. After being disconnected from the power source, it is connected across a 7-MΩ resistor to discharge.

(A) What is the initial discharge current and
(B) How long will it take for the capacitor voltage to decrease to 37 percent of the 20 kV?

**390.** A charged capacitor is connected across a 10-kΩ resistor and allowed to discharge. The potential difference across the capacitor drops to 0.37 of its original value after a time of 7 s. What is the capacitance of the capacitor?

**391.** Describe the current buildup process for an *R-L* circuit (see Figure 24.2) and the associated time constant.

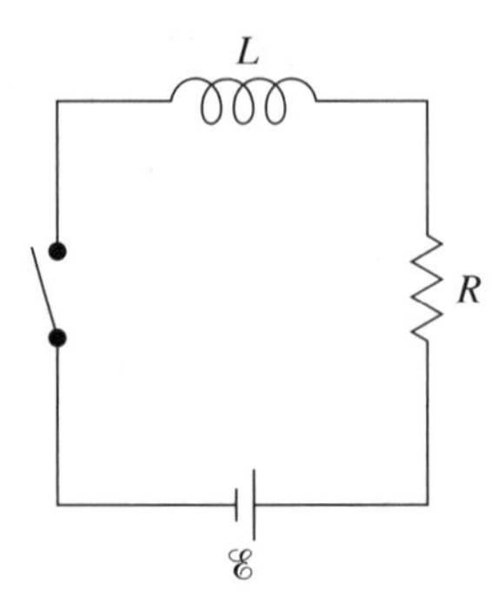


**Figure 24.2**

**392.** A coil has an inductance of 1.5 H and a resistance of 0.6 Ω. If the coil is suddenly connected across a 12-V battery, find the time required for the current to rise to 0.63 of its final value. What will be the final current through the coil?

**393.** A constant potential difference of 60 V is suddenly applied to a coil that has a resistance of 30 Ω and a self-inductance of 8 mH.

(A) At what rate does the current begin to rise?
(B) What is the current at the instant the rate of change of current is 500 A/s?
(C) What is the final current?

CHAPTER 25

# Light and Optical Phenomena

## Reflection and Refraction

**394.** Light having a free-space wavelength of $\lambda = 500$ passes from vacuum into diamond ($n_d = 2.4$). Compute the wave's speed and wavelength in the diamond.

**395.** (A) What is the speed of light in crystalline quartz, whose index of refraction is 1.53?
(B) The speed of light in zircon is $1.52 \times 10^8$ m/s. What is the index of refraction of zircon?

**396.** (A) When light enters a medium of index of refraction $n$, its frequency does not change, but its wavelength and speed do. Show that the wavelength $\lambda'$ in the medium is $\lambda' = \lambda / n$ where $\lambda$ is the wavelength of the light in vacuum.
(B) What is the wavelength in water of a blue light whose wavelength in air is 420 nm? ($n = 1.33$ for water)

**397.** Light from a sodium lamp ($\lambda_0 = 589$ nm) passes through a tank of glycerin (refractive index 1.47) 20 m long in a time $t_1$. If it takes a time $t_2$ to traverse the same tank when filled with carbon disulfide (index 1.63), determine the difference $t_2 - t_1$.

**398.** At what angle must a ray of light be incident on acetone to be refracted into the liquid at 25°? ($n = 1.36$)

**399.** (A) What is the speed of light in water?
(B) Find the angle of refraction of light incident on a water surface at an angle of 48° to the normal. ($n = 1.33$)

**400.** The index of refraction of $n$-propyl alcohol is 1.39. Find the speed of light in that medium and the angle of refraction if light comes from air with an angle of incidence of 55°.

**401.** Light passes from air into a liquid and is deviated 19° when the angle of incidence is 52°. What is the index of refraction of the liquid?

**402.** The index of refraction of a glass sphere is $n_1 = 1.76$. For rays originating within the sphere, find the critical angle if the sphere is immersed in

(A) air ($n_2 = 1$)
(B) water ($n_2 = 1.33$)

**403.** When a fish looks up at the surface of a perfectly smooth lake, the surface appears dark except inside a circular area directly above it. Calculate the angle $\phi$ that this illuminated region subtends.

**404.** A beam of sodium light passes from air into water and then into flat glass, all with parallel surfaces. If the angle of incidence in the air is 45°, what is the angle of refraction in the glass? ($n = 1.33$ and 1.63 for water and flint glass, respectively)

**405.** A layer of benzene (index of refraction = 1.50) floats on water. If the angle of incidence of the light entering the benzene from air is 60°, what is the angle the light makes with the vertical in the benzene and in the water?

**406.** A beam of light is passing through air to oil to water.

(A) If the angle of incidence in the air is 40°, find the angle of refraction in the water. The index of refraction for the oil is 1.45.
(B) If possible, find the angle of incidence in air such that the beam will not enter the water.
(C) Suppose that the direction of the beam is reversed. If possible, find the angle of incidence in water so that the beam will not enter the air.

**407.** A narrow beam of light strikes a glass plate ($n = 1.60$) at an angle of 53° to the normal (see Figure 25.1). If the plate is 20 mm thick, what will be the lateral displacement CE of the beam after it emerges from the plate?

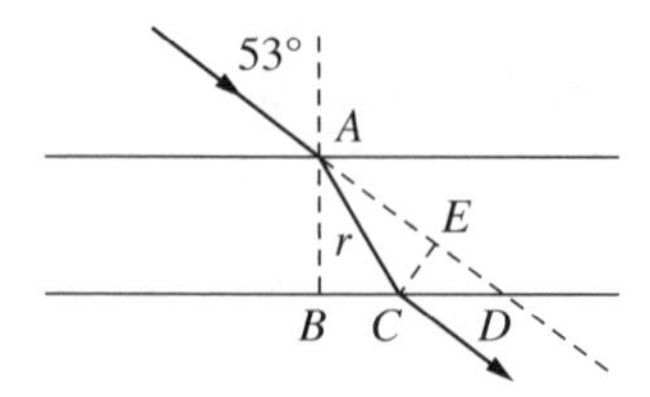


**Figure 25.1**

**408.** Light enters a glass prism having a refracting angle of 60° (see Figure 25.2). If the angle of incidence is 30° (incident ray is parallel to base) and the index of refraction of the glass is 1.50, what is the angle the ray leaving the prism makes with the normal?

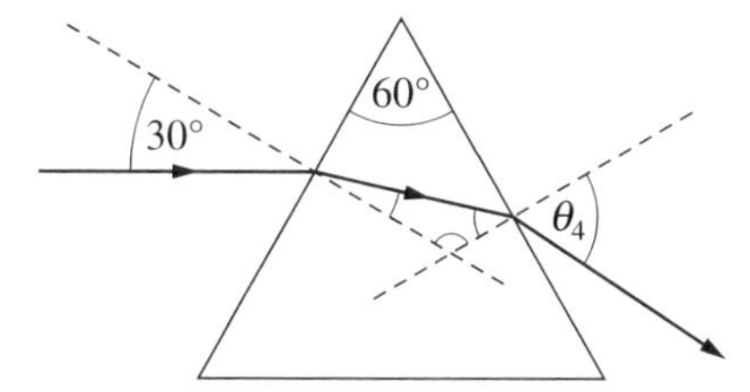


**Figure 25.2**

**409.** The index of refraction of heavy flint glass is 1.68 at 434 nm and 1.65 at 6.71 nm. Calculate the difference in the angle of deviation of blue (434 nm) and red (671 nm) light incident at 65° on one side of a heavy-flint-glass prism with apex angle 60°.

CHAPTER 26

# Mirrors, Lenses, and Optical Instruments

## Mirrors

**410.** State the sign and measurement conventions governing thin lenses and thin spherical mirrors.

**411.** Describe the image formed by a concave mirror when a real object is situated outside the center of curvature, *C*.

**412.** Describe the image formed by a concave mirror when a real object is situated inside the focal point, *F*.

**413.** A concave mirror has a radius of curvature of 0.80 m. Where does this mirror bring sunlight to a focus?

**414.** An object is placed 0.15 m from a concave mirror of focal length 0.20 m.

(A) Where is the image produced?
(B) If the object is 10 cm high, how high is the image?

**415.** An object 10 cm high is 50 cm from a concave mirror of 20 cm focal length. Find the image distance, height, and orientation.

**416.** From the mirror formula, determine where an object must be placed if the image in a concave mirror of focal length $f$ is to be the same size as the object.

**417.** An object is 0.5 ft in front of a concave mirror, and the image is located 2.0 ft behind the mirror. Find the focal length and the radius of curvature of the mirror.

**418.** How far should an object be from a concave spherical mirror of radius 36 cm to form a real image one-ninth its size?

**419.** In a haunted house, it is desired to cast the image of a lamp, magnified five times, upon a wall 12 m distant from the lamp. What kind of spherical mirror is required and what is its position?

**420.** An object 12 mm high is placed 0.5 m from a concave mirror with a radius of curvature of 0.2 m. Find the focal length and the location, height, and orientation of the image.

**421.** As the position of an object reflected in a concave shell mirror of 0.25 m focal length is varied, the position of the image varies. Plot the image distance as a function of the object distance, letting the latter change from 0 to $+\infty$. Where is the image real? Where virtual?

**422.** An object 8 mm high is located 125 mm in front of a concave mirror that has a focal length of 200 mm. Find the position, size, and character of the image.

**423.** What magnification will be obtained by using a concave mirror with a focal length 18 in, if the mirror is held 12 in from an object?

**424.** A man is shaving with his chin 0.4 m from a concave magnifying mirror. If the magnification is 2.5, what is the radius of curvature of the mirror?

**425.** To give a magnified image of a cavity, a dentist holds a small mirror with a focal length of 12 mm a distance of 9 mm from a tooth. What is the magnification obtained?

**426.** An object is placed in front of a concave mirror having a radius of curvature of 0.3 m. How far from the mirror should an object be placed in order to produce

(A) a real image 3 times as large as the object
(B) a virtual image 3 times as large as the object

**427.** An object is 375 mm from a concave mirror of 250-mm focal length.

(A) Find the image distance.
(B) If the object is moved 5 mm farther from the mirror, how far does the image move?

**428.** A convex mirror has a radius of curvature of 90 cm.

(A) Where is the image of an object 70 cm from the mirror formed?
(B) What is the magnification?

**429.** The magnitude of the focal length of a convex mirror is 12 cm. If an object 6 cm high is placed 24 cm from the mirror, what will the image distance be?

**430.** What is the focal length of a convex spherical mirror that produces an image one-sixth the size of an object located 12 cm from the mirror?

**431.** A man's eye is 175 mm from the center of a spherical, reflecting Christmas tree ornament that is 100 mm in diameter. Find the position of the image of his eye, and the magnification.

**432.** An object 28 mm high is 0.48 m from a convex mirror with a radius of curvature of 0.32 m.

(A) Locate the image.
(B) Is the image real or virtual?
(C) Is the image erect or inverted? What is the size of the image?

**433.** A fortune-teller uses a polished sphere of 8-in radius.

(A) If her eye is 10 in from the sphere, where is the image of the eye?
(B) Where would her eye have to be for the image to be at the back surface of the sphere?

**434.** Describe the image formed by a thin converging lens when a real object is situated *outside* the front focal point, *F*.

**435.** Repeat Problem 434 for a real object *inside F*.

**436.** Describe the image of a real object formed by a thin diverging lens.

**437.** A converging lens images the sun at a distance of 20 cm from the lens.

(A) What is the focal length of the lens?
(B) If an object is placed 100 cm from the lens, where will its image be formed? Is it real?
(C) Recalculate if the object is placed 25 cm from the lens.
(D) Recalculate if the object is placed 10 cm from the lens.

**438.** An object is placed 30 cm from a converging lens with 10-cm focal length.

(A) Find the position of the image. Is it real or virtual? Erect or inverted?
(B) Repeat for a 5-cm object distance.

**439.** Repeat Question 438 if the lens is diverging.

**440.** Where must an object be placed in the case of a converging lens of focal length $f$ if the image is to be virtual and three times as large as the object?

**441.** A diverging lens forms an image one-third the size of an object that is 24 cm from the lens. Determine the focal length of the lens.

**442.** A double-convex lens is used to project a slide. The slide is 2 in high and 10 in from the lens. The image is 90 in from the lens. What is the focal length of the lens?

**443.** An object is 6 cm high and 30 cm from a double-convex lens. Its image is 90 cm from the lens and on the same side as the object. What is the focal length of the lens?

**444.** Compute the position and focal length of the converging lens that will project the image of a lamp, magnified 4 diameters, upon a screen 10 m from the lamp.

**445.** A slide projector has a projecting lens of focal length 20 cm. If the slide is 25 cm from this lens, what is the distance to the screen for a clear image?

**446.** A slide projector has a lens whose focal length is 15 cm. If a 5-cm slide is to appear 1 m by 1 m when projected, how far from the lens should the screen be placed?

CHAPTER 27

# Interference, Diffraction, and Polarization

## Interference of Light

**447.** The interference pattern of two identical slits separated by a distance $d = 0.25$ mm is observed on a screen at a distance of 1 m from the plane of the slits. The slits are illuminated by monochromatic light of wavelength 589.3 nm (sodium D) traveling perpendicular to the plane of the slits. Bright bands are observed on each side of the central maximum. Calculate the separation between adjacent bright bands.

**448.** Light from a sodium vapor lamp (589 nm) forms an interference pattern on a screen 0.8 m from a pair of slits. The bright fringes in the pattern are 0.35 cm apart. What is the slit separation?

**449.** Laser light (630 nm) incident on a pair of slits produces an interference pattern in which the bright fringes are separated by 8.3 mm. A second light produces an interference pattern in which the bright fringes are separated by 7.6 mm. What is the wavelength of this second light?

**450.** In Young's interference experiment, two slits are illuminated with orange light of wavelength 6000 Å. The interference pattern is observed on a screen very far from the slits. If the central bright fringe is numbered zero, what must be the path difference for the light from the two slits at the fourth bright fringe?

**451.** Two wavelengths, $\lambda_1$ and $\lambda_2$, are used in the double-slit experiment. If one is 430 nm, what value must the other have for the fourth-order bright fringe of one to fall on the sixth-order bright fringe of the other?

**452.** A soap film has an index of refraction of 1.33. What is the smallest thickness of this film that will give an interference maximum when light of wavelength $\lambda = 500$ nm is incident normally upon it?

**453.** As shown in Figure 27.1, two flat glass plates touch at one edge and are separated at the other edge by a spacer. Using vertical viewing and light with $\lambda = 589$ nm, five dark fringes (D) are obtained from edge to edge. What is the thickness of the spacer?

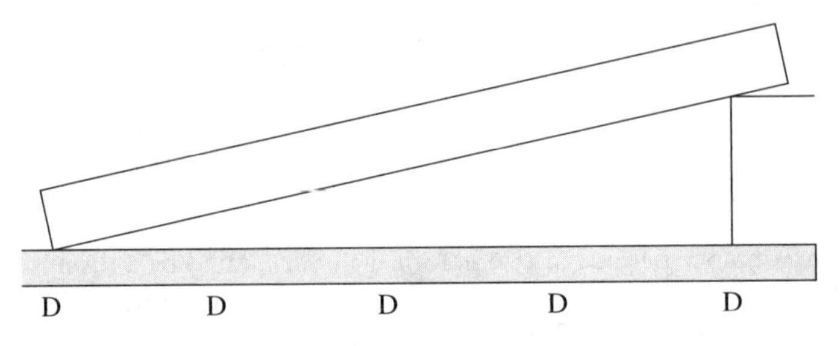


**Figure 27.1**

**454.** A diffraction grating is ruled with 6000 lines per centimeter. The first order of a spectral line is observed to be diffracted at an angle of 30°. What is the wavelength of this radiation?

**455.** A grating having 15,000 lines per inch produces spectra of a mercury arc. The green line of the mercury spectrum has a wavelength of 5461 Å. What is the angular separation between the first-order green line and the second-order green line?

**456.** A light source emits a mixture of wavelengths from 450 to 600 nm. When a diffraction grating is illuminated normally by this source, it is noted that two adjacent spectra barely overlap at an angle of 30°. How many lines per meter are ruled on the grating?

**457.** If a beam of polarized light has one-tenth of its initial intensity after passing through an analyzer, what is the angle between the axis of the analyzer and the initial amplitude of the beam?

**458.** The amplitude of a beam of polarized light makes an angle of 65° with the axis of a Polaroid sheet. What fraction of the beam is transmitted through the sheet?

**459.** Unpolarized light of intensity $I'$ is incident upon a stack of two filters whose transmission axes make an angle $\theta$. Express the intensity $I$ of the emerging beam.

**460.** (A) Ordinary light incident on one polarizing sheet falls on a second polarizing whose plane of vibration makes an angle of 30° with that of the first polarizing sheet. If the polarizing sheets are assumed to be ideal, what is the fraction of the original light transmitted through both polarizing sheets?

(B) If the second polarizing sheet is rotated until the transmitted intensity is 10 percent of the incident intensity, what is the new angle?

**461.** Polarized light of initial intensity $I_0$ passes through two analyzers—the first with its axis at 45° to the amplitude of the initial beam and the second with its axis at 90° to the initial amplitude (Figure 27.2). What is the intensity of the light that emerges from this system, and what is the direction of its amplitude?

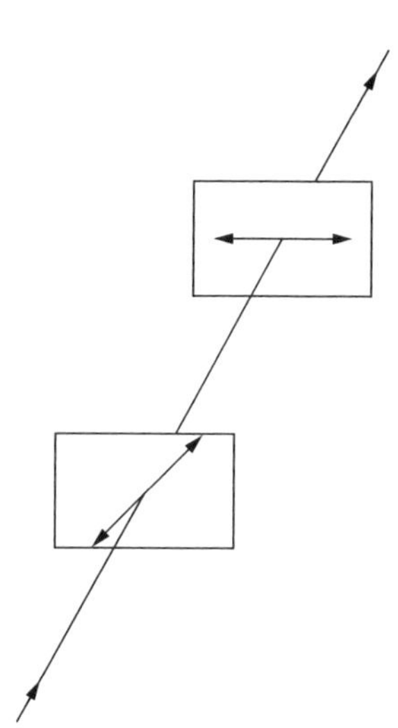

**Figure 27.2**

CHAPTER 28

# Special Relativity

**462.** A spaceship of (rest) length 100 m takes 4 $\mu$s to pass an observer on earth. What is its velocity relative to the earth?

**463.** A beam of radioactive particles is measured as it shoots through the laboratory. It is found that, on the average, each particle "lives" for a time of 20 ns; after that time, the particle changes to a new form. When at rest in the laboratory, the same particles "live" 7.5 ns on the average. How fast are the particles in the beam moving?

**464.** A certain strain of bacteria doubles in number each 20 days. Two of these bacteria are placed on a spaceship and sent away from the earth for 1000 earth days. During this time, the speed of the ship was $0.9950c$. How many bacteria would be aboard when the ship lands on the earth?

## Lorentz Transformation, Length Contraction, Time Dilation, and Velocity Transformation

**465.** A person in a spaceship holds a meterstick as the ship shoots past the earth with a speed $v$ parallel to the earth's surface. What does the person in the ship notice as the stick is rotated from parallel to perpendicular to the ship's motion?

**466.** The insignia painted on the side of a spaceship is a circle with a line across it at 45° to the vertical. As the ship shoots past another ship in space, with a relative speed of $0.95c$, the second ship observes the insignia. What angle does the observed line make to the vertical?

**467.** Rocket $A$ travels to the right and rocket $B$ travels to the left, with velocities $0.8c$ and $0.6c$, respectively, relative to the earth. What is the velocity of rocket $A$ measured from rocket $B$?

**468.** At what speeds will the galilean and Lorentz expressions for $u_x'$ differ by 2 percent?

**469.** As a rocket ship sweeps past the earth with speed $v$, it sends out a pulse of light ahead of it. How fast does the light pulse move according to people on earth?

## Mass-Energy Relation; Relativistic Dynamics

**470.** A 2000-kg car is moving at 15 m/s. How much larger than its rest mass is its mass at this speed? (*Hint*: For $x$ very small, $1/\sqrt{1-x} \approx 1+\frac{1}{2}x$.)

**471.** Determine the mass and speed of an electron having kinetic energy of 100 keV ($1.6 \times 10^{-14}$ J).

**472.** Find the speed that a proton must be given if its mass is to be twice its rest mass of $1.67 \times 10^{-27}$ kg. What energy must be given the proton to achieve this speed?

**473.** How much energy must be given to an electron to accelerate it to $0.95c$?

**474.** A 2-kg object is lifted from the floor to a tabletop 30 cm above the floor. By how much did the mass of the object increase because of its increased PE?

**475.** Find the increase in mass of 100 kg of copper ($c = 0.389$ kJ/kg · K) if its temperature is increased 100°C.

**476.** If 1 g of matter could be converted entirely into energy, what would be the value of the energy so produced at \$0.01 per kW · h?

CHAPTER 29

# Particles of Light and Waves of Matter

## Photons and the Photoelectric Effect

**477.** A certain radio station transmits at a frequency of 900 kHz. What is the wavelength of its waves? An observing tower is located 50 km from the radio station. How many wave crests pass the observing tower each second?

**478.** Find the energy of the photons in a beam whose wavelength is 526 nm.

**479.** What is the photon energy in joules corresponding to a 60-Hz wave emitted from a power line? How does this compare with the energy range for light?

**480.** Imagine a source emitting 100 W of green light of a wavelength of 500 nm. How many photons per second are emerging from the source?

**481.** A sensor is exposed for 0.1 s to a 200-W lamp 10 m away. The sensor has an opening that is 20 nm in diameter. How many photons enter the sensor if the wavelength of the light is 600 nm? Assume that all the energy of the lamp is given off as light.

**482.** What wavelength must electromagnetic radiation have if a photon in the beam is to have the same momentum as an electron moving with a speed $2 \times 10^5$ m/s?

**483.** How many red photons ($\lambda$ = 663 nm) must strike a totally reflecting screen per second, at normal incidence, if the exerted force is to be 1 N?

**484.** What potential difference must be applied to stop the fastest photoelectrons emitted by a nickel surface under the action of ultraviolet light of wavelength 2000 Å? The work function of nickel is 5.00 eV.

**485.** The work function of sodium metal is 2.3 eV. What is the longest wavelength light that can cause photoelectron emission from sodium?

**486.** Will photoelectrons be emitted by a copper surface, of work function 4.4 eV, when illuminated by visible light?

**487.** Light of wavelength 600 nm falls on a metal having photoelectric work function 2 eV. Find

(A) the energy of a photon
(B) the kinetic energy of the most energetic photoelectron
(C) the stopping potential

**488.** A photon of energy 4.0 eV imparts all its energy to an electron that leaves a metal surface with 1.1 eV of kinetic energy. What is the work function $W_{min}$ of the metal?

## Compton Scattering; X-Rays; Pair Production and Annihilation

**489.** It is proposed to send a beam of electrons through a diffraction grating with the distance between slits being *d*. The electrons have a speed of 400 m/s. How large must *d* be if a strong beam of electrons is to emerge at an angle of 25° to the straight-through beam?

CHAPTER 30

# Modern Physics: Atoms, Nuclei, and Solid-State Electronics

## Atoms and Molecules

**490.** Sodium atoms emit a spectral line with a wavelength in the yellow, 589.6 nm. What is the difference in energy between the two energy levels involved in the emission of this spectral line?

**491.** X-rays of wavelength 1.37 nm incident on an atom cause photoemission of its electrons. If the emitted electrons have an energy of 83 eV, what is the energy of the level from which the electrons were ejected?

## Nuclei and Radioactivity

**492.** What is the number of neutrons in the nucleus of $^{23}_{11}\text{Na}$?

**493.** What is the binding energy of $^{12}\text{C}$?

**494.** In a particular fission reaction, a $^{235}_{92}\text{U}$ nucleus captures a slow neutron; the fission products are three neutrons, a $^{142}_{57}\text{La}$ nucleus, and a fission product $_Z\text{X}$. Determine $Z$.

**495.** When an atom of $^{235}\text{U}$ undergoes fission in a reactor, about 200 MeV of energy is liberated. Suppose that a reactor using uranium-235 has an output of 700 MW and is 20 percent efficient.

(A) How many uranium atoms does it consume in one day?
(B) What mass of uranium does it consume each day?

**496.** Neutrons are frequently detected by allowing them to be captured by boron-10 nuclei in a counter (similar to a Geiger counter) filled with $B^{10}F_3$ gas. The reaction is

$$ {}_0^1n + {}_5^{10}B \rightarrow {}_3^7Li + {}_2^4He $$

Compute the energy released in this reaction. The relevant rest masses are neutron, 1.00866 u; boron-10, 10.01294 u; lithium-7, 7.01600 u; helium-4, 4.00260 u.

**497.** Complete and balance the equations for the following nuclear reactions by replacing the question mark with the correct symbol.

(A) ${}_{86}^{222}Rn \rightarrow ? + {}_2^4He$

(B) ${}_1^2H + {}_1^2H \rightarrow ? + {}_1^1H$

(C) ${}_{93}^{239}Np \rightarrow ? + {}_{-1}^{0}\beta$ (electron)

(D) ${}_{11}^{22}Na \rightarrow ? + {}_{+1}^{0}\beta$ (positron)

**498.** Complete the following nuclear equations.

(A) ${}_7^{14}N + {}_2^4He \rightarrow {}_8^{17}O + ?$

(B) ${}_4^9Be + {}_2^4He \rightarrow {}_6^{12}C + ?$

**499.** Complete the following nuclear equations.

(A) ${}_{15}^{30}P \rightarrow {}_{14}^{30}Si + ?$

(B) ${}_1^3H \rightarrow {}_2^3He + ?$

**500.** Cobalt-60 ($^{60}Co$) is often used as a radiation source in medicine. It has a half-life of 5.25 years. How long, after a new sample is delivered, will the activity have decreased

(A) to about one-eighth its original value
(B) to about one-third its original value

# ANSWERS

## Chapter 1: Equilibrium of Concurrent Forces

**1.** A scalar quantity has only magnitude; it is a pure number, positive or negative. Scalars, being simple numbers, are added, subtracted, and so on, in the usual way. It may have a unit after it, e.g., mass = 3 kg.

**2.** A vector quantity has both magnitude and direction. For example, a car moving south at 40 km/h has a *vector velocity* of 40 km/h southward.

A vector quantity can be represented by an arrow drawn to scale. The length of the arrow is proportional to the magnitude of the vector quantity (40 km/h in the above example). The direction of the arrow represents the direction of the vector quantity.

**3.** The method for finding the resultant of several vectors consists of beginning at any convenient point and drawing (to scale) each vector arrow in turn. They may be taken in any order of succession. The tail end of each arrow is attached to the tip end of the preceding one.

The resultant is represented by an arrow with its tail end at the starting point and its tip end at the tip of the last vector added.

**4.** A component of a vector is its "shadow" (perpendicular drop) on an axis in a given direction. For example, the *p*-component of a displacement is the distance along the *p* axis corresponding to the given displacement. It is a scalar quantity, being positive or negative, as it is positively or negatively directed along the axis in question. In Figure A1.1, $A_p$ is positive. (One sometimes defines a vector component as a *vector* pointing along the axis and having the size of the scalar component. If the scalar component is negative, the vector component points in the negative direction along the axis.)

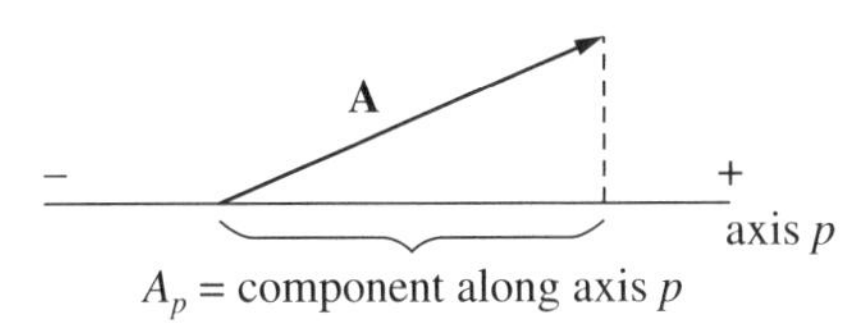


**Figure A1.1**

**5.** Two forces act upon the object: the upward pull of the cord and the downward pull of gravity. Represent the pull of the cord by $T$, the tension in the cord. The pull of gravity, the weight of the object, is $w$ = 50 N. These two forces are shown in the free-body diagram, Figure A1.2.

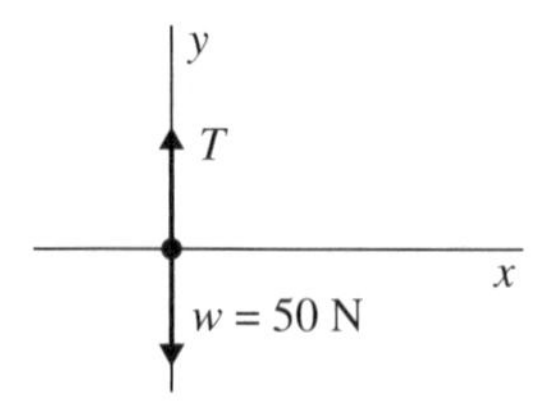


**Figure A1.2**

The forces are already in component form, so we can write the first condition for equilibrium at once,

$$\sum F_x = 0 \quad \text{becomes} \quad 0 = 1$$
$$\sum F_y = 0 \quad \text{becomes} \quad T - 50\text{ N} = 0$$

from which $T = 50$ N.

**6.**

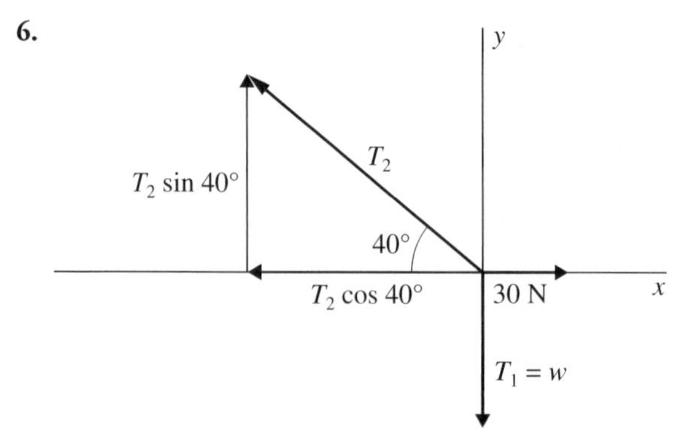


**Figure A1.3**

As seen in Question 5, the tension in cord 1 is equal to the weight of the object hanging from it. Therefore, $T_1 = w$, and we wish to find $T_1$ or $w$.

Note that the unknown force, $T_1$, and the known force, 30 N, both pull on the knot at point $P$. It therefore makes sense to isolate the knot at $P$ as our object. The free-body diagram showing the forces on the knot is drawn as Figure A1.3. The force components are also found there.

Next, write the first condition for equilibrium for the knot. From the free-body diagram,

$$\sum F_x = 0 \quad \text{becomes} \quad 30\text{ N} - T_2 \cos 40° = 0$$
$$\sum F_y = 0 \quad \text{becomes} \quad T_2 \sin 40° - w = 0$$

Solving the first equation for $T_2$ gives $T_2 = 39$ N. Substituting this value in the second equation gives $w = 25$ N as the weight of the object.

**7.** $B$ will experience the largest tension, 200 N in this case. Solve for forces along the vertical: $W = 200 \sin 53° = 160$ N, and in the horizontal direction, find $A = 200 \cos 53° = 120$ N.

**8.** Let us select as our object the knot at $A$ because we know one force acting on it. The weight pulls down on it with a force of 600 N, and so the free-body diagram for the knot. Applying the first condition for equilibrium to that diagram, we have

$$\sum F_x = 0 \quad \text{or} \quad T_2 \cos 60° - T_1 \cos 60° = 0$$

$$\sum F_y = 0 \quad \text{or} \quad T_1 \sin 60° + T_2 \sin 60° - 600 = 0$$

The first equation yields $T_1 = T_2$. Substitution of $T_1$ for $T_2$ in the second equation gives $T_1 = 346$ N, and this is also $T_2$.

Let us now isolate knot $B$ as our object. Its free-body diagram is shown in Figure 2.11($b$). We have already found that $T_2 = 346$ N, and so the equilibrium equations are

$$\sum F_x = 0 \quad \text{or} \quad T_3 \cos 20° - T_5 - 346 \sin 30° = 0$$

$$\sum F_y = 0 \quad \text{or} \quad T_3 \sin 20° - 346 \cos 30° = 0$$

The last equation yields $T_3 = 877$ N. Substituting this in the prior equation gives $T_5 = 651$ N.

We can now proceed to the knot at $C$ and the free-body diagram of Figure 2.11($c$). Recalling that $T_1 = 346$ N,

$$\sum F_x = 0 \quad \text{becomes} \quad T_5 + 346 \sin 30° - T_4 \cos 20° = 0$$

$$\sum F_y = 0 \quad \text{becomes} \quad T_4 \sin 20° - 346 \cos 30° = 0$$

The latter equation yields $T_4 = 877$ N.
(Note that from the symmetry of the system we could have deduced $T_1 = T_2$ and $T_4 = T_3$.)

**9.**

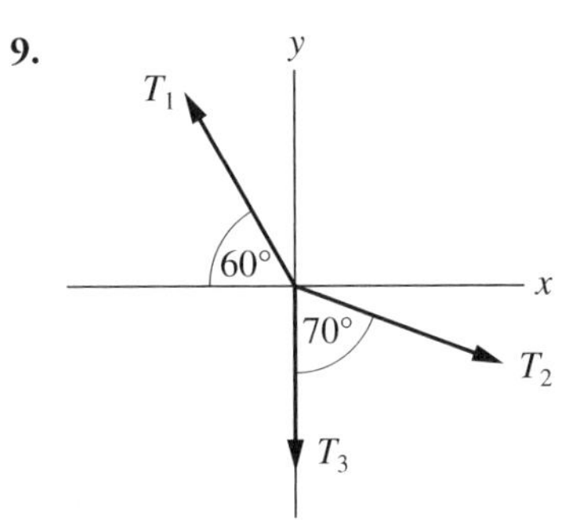


**Figure A1.4**

The knot is in equilibrium under the action of three forces, and the free-body diagram is as shown in Figure A1.4. $T_3 = w = 40$ N.

$$\sum F_x = 0 \Rightarrow T_2 \sin 70° - T_1 \cos 60° = 0 \quad \text{or} \quad (0.940)T_2 = (0.500)T_1, \quad T_1 = 1.88T_2$$

$$\sum F_y = 0 \Rightarrow T_1 \sin 60° - T_2 \cos 70° - T_3 = 0, \quad (0.866)T_1 - (0.342)T_2 = T_3 = 40 \text{ N}$$

Substituting for $T_1$,

$$(0.866)(1.88T_2) - (0.342)T_2 = 40 \text{ N} \quad 1.29T_2 = 40 \text{ N} \quad T_2 = 31 \text{ N} \quad \text{and}$$
$$T_1 = (1.88)(31.0) = 58 \text{ N}$$

**10.**

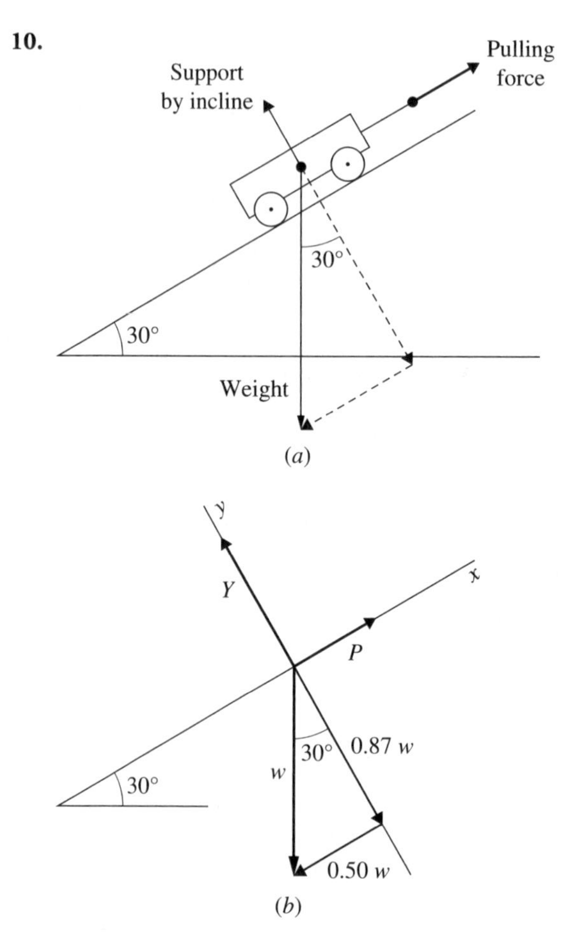


**Figure A1.5**

The situation is shown in Figure A1.5(*a*). Because the wagon moves at constant speed along a straight line, its velocity vector is constant. Therefore, the wagon is in translational equilibrium, and the first condition for equilibrium applies to it.

We isolate the wagon as the object. Three forces act on it: (1) the pull of gravity $w$ (its weight), directed straight down; (2) the force $P$ exerted on the wagon parallel to the incline to pull it up the incline; and (3) the push $Y$ of the incline that supports the wagon. These three forces are shown in the free-body diagram, Figure A1.5(*b*).

For situations involving inclines, it is convenient to take the $x$ axis parallel to the incline and the $y$ axis perpendicular to it. After taking components along these axes, we can write the first condition for equilibrium.

$$\sum F_x = 0 \quad \text{becomes} \quad P - 0.50w = 0 \quad \sum F_y = 0 \quad \text{becomes} \quad Y - 0.87w = 0$$

Solving the first equation and recalling that $w = 200$ N, we find that $P = 0.50w = 100$ N. The required pulling force is 100 N.

**11.**

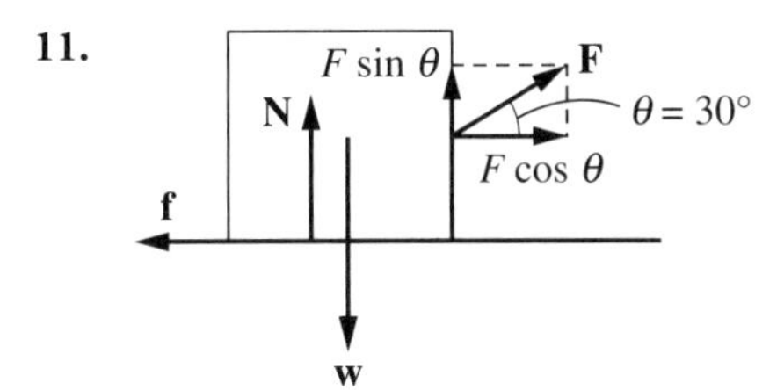


**Figure A1.6**

First draw a force diagram, Figure A1.6. Next, consider the force in the $x$ direction and apply the conditions for equilibrium, noting $f$ equals its maximum value to start motion.

$$\sum F_x = 0 \quad F \cos\theta - f = 0 \quad F \cos\theta = f \quad 0.866F = f = \mu_s N = 0.4N$$

Now apply the conditions for equilibrium to the forces in $y$ direction.

$$\sum F_y = 0 \quad N + F \sin\theta - W = 0 \quad N + 0.5F - 100 = 0 \quad N = 100 - 0.5F$$

Substituting this equation for $N$ in $0.866F = 0.4N$ above,

$$0.866F = 0.4(100 - 0.5F) \quad 0.866F + 0.2F = 40 \quad F = 38 \text{ N}$$

**12.**

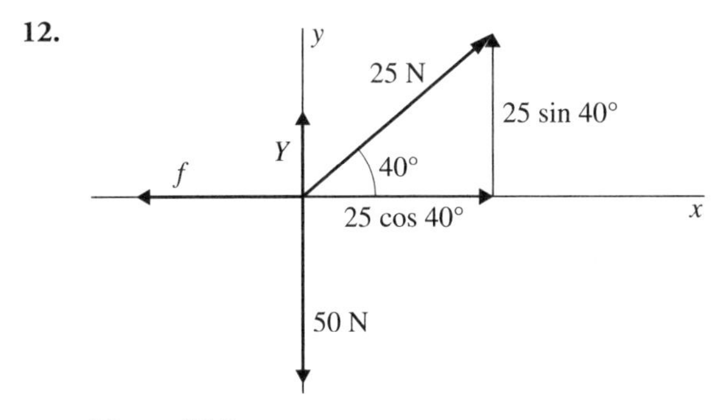


**Figure A1.7**

Note the forces acting on the box, as shown in Figure A1.7. The friction is $f$, and the normal force, the supporting force exerted by the floor, is $Y$. The free-body diagram and components are shown in Figure A1.7. Because the box is moving with constant velocity, it is in equilibrium. The first condition for equilibrium tells us that

$$\sum F_x = 0 \quad \text{or} \quad 25 \cos 40° - f = 0$$

We can solve for $f$ at once to find that $f = 19$ N. The friction force is 19 N. To find $Y$, we use the fact that

$$\sum F_y = 0 \quad \text{or} \quad Y + 25 \sin 40° - 50 = 0$$

Solving gives the normal force as $Y = 34$ N.

**13.** We apply $\sum F_y = 0$ in each case.

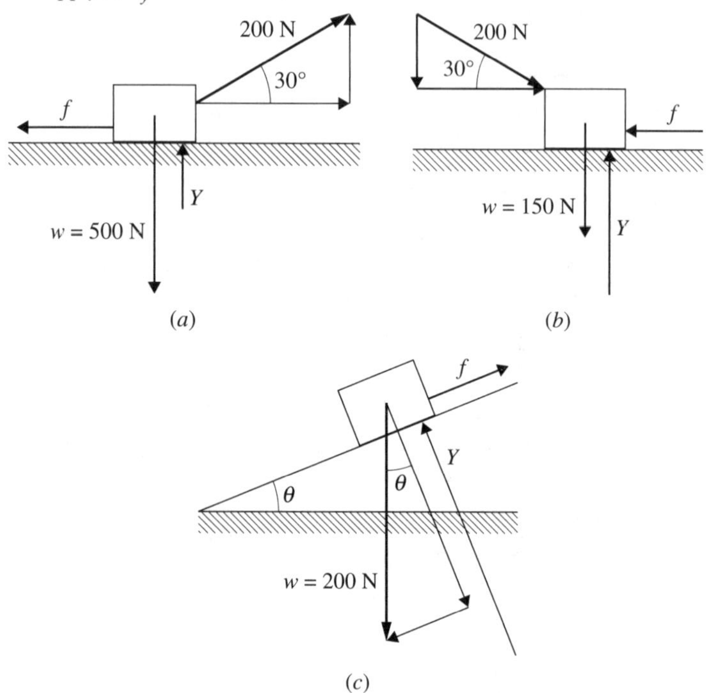


**Figure A1.8**

**(A)** $Y + 200 \sin 30° - 500 = 0$ from which $Y = 400$ N
**(B)** $Y - 200 \sin 30° - 150 = 0$ from which $Y = 250$ N
**(C)** $Y - 200 \cos \theta = 0$ from which $Y = (200 \cos \theta)$ N

**14.**

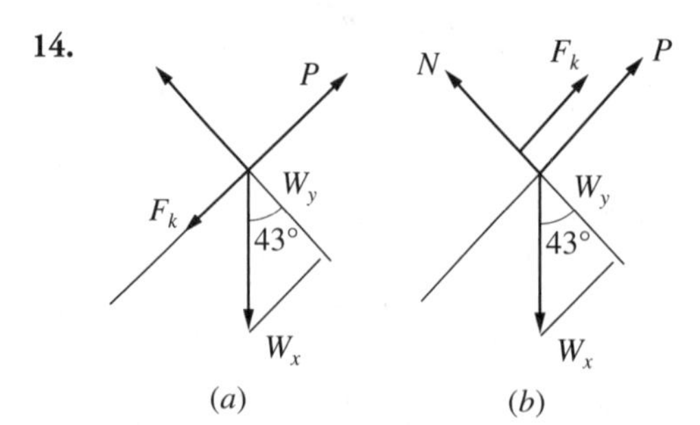


**Figure A1.9**

The weight components are

$$W_x = (60\text{ N})\sin 43^\circ = 41\text{ lb} \qquad W_y = (60\text{ N})\cos 43^\circ = 44\text{ N}$$

**(A)** Using Figure A1-9(*a*),

$$\sum F_y = 0 \quad N - W_y = 0 \quad N = 49\text{ N}$$

$$F_k = \mu_k N = 0.3(43.88) \quad F_k = 13\text{ N}$$

$$\sum F_x = 0 \quad P - F_k - W_x = 0 \quad P = 13.16 + 40.92 \quad P = 54\text{ N}$$

**(B)** The push is now just enough to keep the motion plane at constant speed. From Figure A1.9(*b*),

$$\sum F_x = 0 \quad P - F_k - W_x = 0 \quad P = 40.92\text{ N} - 13.16\text{ N} \qquad P = 28\text{ N}$$

**15. (A)** $\sum F_x = 0 \Rightarrow F\cos 55^\circ - f = 0 \quad (N_x = w_x = 0) \quad \text{or} \quad 20(0.573) = f = 11.5\text{ N}$

**(B)** $\mu_k = \dfrac{f}{N} = \dfrac{11.5\text{ N}}{34\text{ N}} = 0.34$

**16. (A)** $\sum F_x = 0 \Rightarrow f - w\sin 40^\circ = 0 \quad f = 60\text{ N}(0.643) = 39\text{ N}$

**(B)** $\mu_k = \dfrac{f}{N} = \dfrac{38.6\text{ N}}{46\text{ N}} = 0.84$

## Chapter 2: Kinematics in One Dimension

**17.** The distance from $x_0$ at the moment fuel is exhausted is $x_1 = (0)(8) + \frac{1}{2}(5)(8)^2 = 160$ m, and at this point $v = (2ax_1)^{1/2} = 40$ m/s. Hence, the distance covered in 12 s is $x_2 = x_2 = x_1 + v(12-8) = 160 + (40)(4) = 320$ m.

**18. (A)** Applying $v = v_0 + at$, $0 = 20 + (-4)t'$ or $t' = 5$ s. Then $x' = v_0 t' + \frac{1}{2}at'^2 = (20)(5) + \frac{1}{2}(-4)(5)^2 = 50$ m. Or, from $v^2 = v_0^2 + 2ax$:

$$0 = (20)^2 + 2(-4)x' \quad \text{or} \quad x' = 50\text{ m.}$$

**(B)** $15 = 20t + \frac{1}{2}(-4)t^2 \quad \text{or} \quad 2t^2 - 20t + 15 = 0$

Solving this quadratic,

$$t = \frac{20 \pm \sqrt{(20)^2 - 4(2)(15)}}{4} = \frac{1}{4}(20 \pm 16.7)$$

Thus, $t_1 = 0.82$ s, $t_2 = 9.17$ s, where $t_1$ is the time from the origin to $x = 15$ m and $t_2$ is the time to go from $O$ out beyond $x = 15$ m and return to that point. At $x = 15$ m,

$$v_1 = 20 - 4(0.82) = +16.7 \text{ m/s} \qquad v_2 = 20 - 4(9.17) = -16.7 \text{ m/s}$$

Observe that the speeds are equal.

**(C)** At $x = 25$ m, $v^2 = (20)^2 + 2(-4)(25)$, or $v = \pm 14.1$ m/s; and at $x = -25$ m, $v^2 = 20^2 + 2\,(-4)(-25)$, or $v = -24.5$ m/s. (Why has the root $v = +24.5$ m/s been discarded?)

Assuming that $x = 55$ m, $v^2 = 20^2 + 2(-4)(55)$, from which $v = \pm\sqrt{-40}$. The imaginary value of $v$ indicates that $x$ never reaches 55 m, as expected from the result of part (*a*).

**19.** **(A)** Choose $y$ downward as positive. Then $a = g = 9.8 \text{ m/s}^2$.

**(B)** For $t = 3.0$ s, $y = v_0 t + \frac{1}{2}at^2 = 0 + \frac{1}{2}(9.8 \text{ m/s}^2)(3.0 \text{ s})^2 = 44$ m.

**(C)** Letting $y = 70$ m, we have

$$v_f^2 = v_0^2 + 2ay = 0 + 2(9.8 \text{ m/s}^2)(70 \text{ m}) = 1372 \text{ m}^2/\text{s}^2 \quad \text{or} \quad v_f = 37 \text{ m/s}$$

**(D)** Letting $v_f$ now equal 25 m/s, we have

$$v_f = v_0 + at \qquad \text{yields} \qquad 25 \text{ m/s} = 0 + (9.8 \text{ m/s}^2)t \quad \text{or} \quad t = 2.55 \text{ s}$$

**(E)** Now we let $y = 300$ m and we have

$$y = v_0 t + \tfrac{1}{2}at^2 \qquad \text{yields} \qquad 300 \text{ m} = 0 + \tfrac{1}{2}(9.8 \text{ m/s}^2)t^2 \quad \text{or} \quad t = 7.8 \text{ s}$$

**20.** **(A)** $v = v_0 + at = 25 + 32(1.5)$ or $v = 73$ ft/s (where $a = g = 32 \text{ ft/s}^2$)

**(B)** For constant acceleration $v_{\text{avg}} = \frac{1}{2}(v + v_0) = \frac{1}{2}(73 + 25) = 49$ ft/s

$$s = v_{\text{avg}}t = 49(1.5) = 73.5 \text{ ft} \quad \text{or} \quad s = v_0 t + \tfrac{1}{2}at^2 = 25(1.5) + \tfrac{1}{2}(32)(1.5)^2$$
$$= 37.5 + 36 = 73.5 \text{ ft}$$

**21.** Choose downward positive; $a = g = 9.8 \text{ m/s}^2$. We are given $v_0 = 8$ m/s.

**(A)** One might directly solve $y = v_0 t + \frac{1}{2}at^2$, or $25 \text{ m} = (8 \text{ m/s})t + \frac{1}{2}(9.8 \text{ m/s}^2)t^2$ for $t$; but it is easier first to find the final velocity.

**(B)** $v_f^2 = v_0^2 + 2ay = (8 \text{ m/s})^2 + 2(9.8 \text{ m/s})(25 \text{ m}) = 554 \text{ m}^2/\text{s}^2$, or $v_f = 23.5$ m/s. Returning to (A), $v_f = v_0 + at$ yields $23.5 \text{ m/s} = 8 \text{ m/s} + (9.8 \text{ m/s}^2)t$, or $t = 1.58$ s.

**22.** Let us take *up* as positive. For the trip from beginning to end, $y = 0$, $a = -9.8 \text{ m/s}^2$, $t = 4$ s. Note that the start and the endpoint for the trip are the same, so the displacement is zero. Use $y = v_0 t + \frac{1}{2}at^2$ to find $0 = v_0(4 \text{ s}) + \frac{1}{2}(-9.8 \text{ m/s}^2)(4 \text{ s})^2$, from which $v_0 = 20$ m/s.

**23.** The initial velocity of the bag when released is the same as that of the balloon, 13 m/s upward. Let us choose *up* as positive and take $y = 0$ at the point of release.

**(A)** At the highest point, $v_f = 0$. From $v_f^2 = v_0^2 + 2ay$, $0 = (13 \text{ m/s})^2 + 2(-9.8 \text{ m/s}^2)y$, or $y = 8.6$ m. The maximum height is $300 + 8.6 = 309$ m.

**(B)** Take the endpoint to be its position at $t = 5$ s. Then, from $y + v_0t + \frac{1}{2}at^2$, $y = (13 \text{ m/s})(5 \text{ s}) + \frac{1}{2}(-9.8 \text{ m/s}^2)(5 \text{ s})^2 = -58$ m. So its height is 300 – 58 = 242 m. Also, from $v_f = v_0 + at$, $v_f = 13 \text{ m/s} + (-9.8 \text{ m/s}^2)(5 \text{ s}) = -36$ m/s. It is moving downward at 36 m/s.

**(C)** Just before it hits the ground, the bag's displacement is –300 m. $y = v_0t + \frac{1}{2}at^2$ becomes $-300 \text{ m} = (13 \text{ m/s})t + \frac{1}{2}(-9.8 \text{ m/s}^2)t^2$, or $4.9t^2 - 13t - 300 = 0$. The quadratic formula gives $t = 9.3$ s (and –6.6 s, which is not physically meaningful).

**24.** On earth we can write $v_E^2 = 2g_E(12)$, while on the moon $v_M^2 = 2g_M h_M$. The throwing velocities are the same, so the second expression can be divided by the first to give $h_M = 12(g_E/g_M) = 12\,(9.80/1.67) = 70$ m.

**25. (A)** $v_0 = 0$, $v = 10^6$ m/s in $x = 10^{-2}$ m of displacement. Then $v^2 = v_0^2 + 2ax$ yields $(10^6 \text{ m/s})^2 = 0 + 2a\,(10^{-2} \text{ m})$, or $a = 5.0 \times 10^{13} \text{ m/s}^2$.

**(B)** $v = v_0 + at$ yields $10^6 \text{ m/s} = 0 + (5.0 \times 10^{13} \text{ m/s}^2)t$, or $t = 2.0 \times 10^{-8}$ s.

**26.** Choose upward as positive for both (A) and (B); $a = -g = -9.8 \text{ m/s}^2$.

**(A)** $v_0 = 0$. To find the height, let $y$ be the displacement at time $t$ (remember the $y = 0$ point is at the balloon), and we are given $t = 20$ s. Then $y = v_0t + \frac{1}{2}at^2 = 0 + \frac{1}{2}(-9.8 \text{ m/s}^2)(20 \text{ s})^2 = -1960$ m. The height is $|y| = 1960$ m.

**(B)** Here the bottle initially has the velocity of the balloon, so $v_0 = 50$ m/s. Now, $y = v_0t + \frac{1}{2}at^2 = (50 \text{ m/s})(20 \text{ s}) + \frac{1}{2}(-9.8 \text{ m/s}^2)(20 \text{ s})^2 = -960$ m. Again the height $= |y| = 960$ m.

**27.** Because the velocity is given by the slope, $\Delta x/\Delta t$, of the tangent line, we take a tangent to the curve at point $A$. The tangent line is the curve itself in this case. For the triangle shown at $A$, we have

$$\frac{\Delta x}{\Delta t} = \frac{4 \text{ m}}{8 \text{ s}} = 0.50 \text{ m/s}$$

This is also the velocity at point $B$ and at every other point on the straight-line graph. It follows that $a = 0$ and $\bar{v}_x = v_x = 0.50$ m/s.

**28.** The tangent at $F$ is the dashed line $GH$. Taking triangle $GHJ$, we have

$$\Delta t = 24 - 4 = 20 \text{ s} \qquad \Delta x = 0 - 15 = -15 \text{ m}$$

Hence, slope at $F$ is $$v_F = \frac{\Delta x}{\Delta t} = \frac{-15 \text{ m}}{20 \text{ s}} = -0.75 \text{ m/s}$$

The negative sign tells us that the object is moving in the $-x$ direction.

**29.** The average velocity is zero, since the displacement vector is zero. The instantaneous velocities are just the slopes of the curve at each point. At $A$ the velocity is 40/6 = 6.7 m/min east. At $B$ it is zero. At $C$ it is –65/5 = –13 m/min east, or +13 m/min west.

**30. (A)** $\bar{v} = (-25 - 40)/(14 - 7) = -9.3$ m/min min east

**(B)** the same as at point $C$, –13 m/min east

**(C)** The slope is +25/(19–14) = 5.0 m/min east. Note that negative velocity east means motion is west.

**31.** See Figure A2.1.

**(A)** $\overline{v} = (4.8 - 0)/(8.0 - 0) = 0.60$ cm/s.

**(B)** From the slope at each point $v = -0.48$ cm/s

**(C)** 1.3 cm/s.

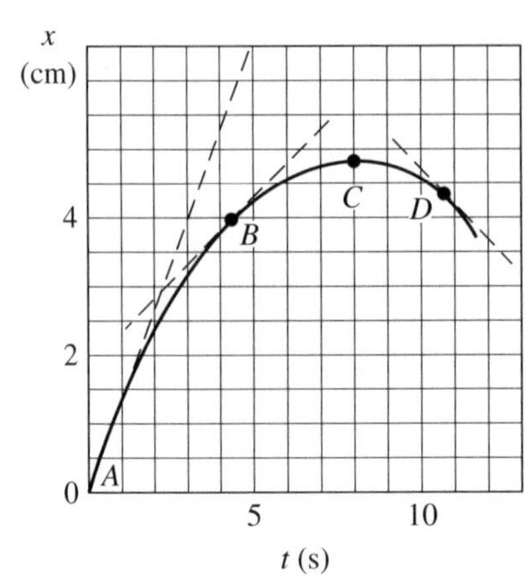


**Figure A2.1**

**32.** See Figure A2.2. The acceleration at any time is the slope of the $v$-vs.-$t$ curve.

**(A)** At $A$ the slope is, from where the tangent line through $A$ cuts the coordinate axes, $a = -7.0/0.73 = -9.6$ m/s$^2$.

**(B)** The slope, and therefore the acceleration, is zero.

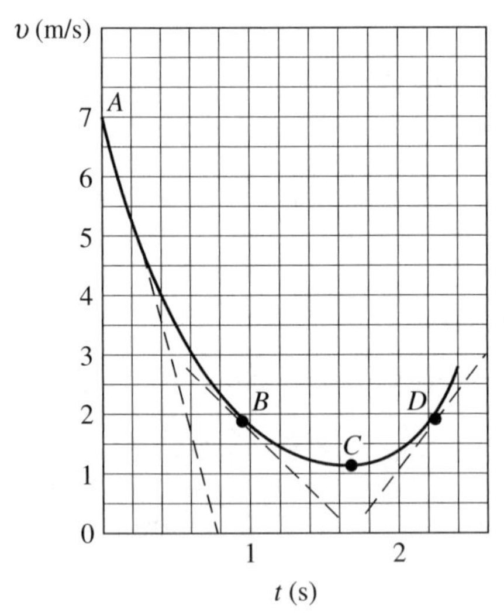


**Figure A2.2**

**33.** **(A)** For the coordinate system shown in Figure 2.7, $y = v_0t + \frac{1}{2}at^2$. But at the edge of the roof $y = 0$, and thus $0 + 20t_1 + \frac{1}{2}(-9.8)t_1^2$, from which $t_1 = 0$, indicating the instant at which the ball is released, and also $t_1 = 4.08$ s, which is the time to go up and return to the edge. Then, from $v = v_0 + at$, $v_1 = 20 + (-9.8)(4.08) = -20$ m/s, which is the negative of the initial velocity.

**(B)** $-50 = 20t_2 + \frac{1}{2}(-9.8)t_2^2$ or $t_2 = 5.8$ s $v_2 = 20 + (-9.8)(5.8) = -37$ m/s

**34. (A)** Maximum height above ground: $h = y_{max} + 50$. From $v_0^2 + 2ay_{max} = 0$,

$$y_{max} = \frac{-(20)^2}{-2(9.8)} = 20 \text{ m}$$

Thus, $h = 70$ m.

**(B)** If $t_1$ and $t_2$ are the times to reach $P_1$ and $P_2$, respectively, $-15 = 20t_1 - 4.9t_1^2$ and $-30 = 20t_2 - 4.9t_2^2$. Solving, $t_1 = 4.723$ s, $t_2 = 5.248$ s, and the time from $P_1$ to $P_2$ is $t_2 - t_1 = 0.53$ s.

**(C)** If $v_0$ is the desired initial velocity, then $-v_0$ is the velocity upon passing the edge. Then, applying $y = v_0t + \frac{1}{2}at^2$ to the trip down the tower, we find $-50 = (-v_0)(3) - 4.9(3)^2$, or $v_0 = 2.0$ m/s.

**35. (A)** At $t = 0$ let the man's position be the origin, $x_{m0} = 0$. The bus door is then at $x_{b0} = 6.0$ m. The equations of motion for the man and the bus are

$$x_m = x_{m0} + v_{m0}t + \tfrac{1}{2}a_m t^2 \qquad x_b = x_{b0} + v_{b0}t + \tfrac{1}{2}a_b t^2$$

Now $\quad v_{m0} = 4.0 \text{ m/s} \quad v_{b0} = 0 \quad a_m = 0 \quad a_b = 1.2 \text{ m/s}^2$

Thus $\quad x_m = 4.0t \quad x_b = 6.0 + 0.6t^2$

When the man catches the bus, $x_m = x_b$, or $4.0t = 6.0 + 0.6t^2$.

This can be reexpressed as $3t^2 - 20t + 30 = 0$. Solving by the quadratic formula

$$t = \frac{20 \pm \sqrt{400 - 360}}{6} = \frac{10 \pm \sqrt{10}}{3} = 2.3 \text{ s}, 4.4 \text{ s}$$

Note that there are two positive time solutions. This can be understood as follows. The first time, $t_1 = 2.3$ s, corresponds to his first reaching the door. This is the real answer to the question. However, the equations we have solved "don't know" he will stop running and board the bus; the equations have him continuing to run at constant speed. He thus goes past the bus, but since the bus is accelerating, it eventually builds up a larger velocity than the man and will catch up with him, $t_2 = 4.4$ s.

**(B)** If the initial position of the bus is 10.0 m, then $x_m = x_b$ yields $3t^2 - 20t + 50 = 0$, which has only complex roots. Thus, no real time exists at which the man catches up.

**36. (A)** The velocities $v_T$ at the top and $v_B$ at the bottom of the window are related by the following equations: $\bar{v} = (v_T + v_B)/2 = 3/0.5 = 6$, so $v_T + v_B = 12$, and $v_B = v_T + g(0.5)$, so $v_B - v_T = 4.9$. Eliminating $v_B$ between these two expressions yields $v_T = 3.55$ m/s.

**(B)** The distance needed to reach this speed is

$$h = \frac{v_T^2}{2g} = \frac{(3.55)^2}{2(9.8)} = 0.64 \text{ m}$$

**37.** The total displacement of the truck from the instant the driver sees the car is $x = v_0\Delta t + x_A$, where $x_A$ is the displacement from the point of deceleration to the point of rest. We are given $v_0 = 21$ m/s and the deceleration is $a = -3$ m/s$^2$. $x_A$ can be obtained from the equation $v_f^2 = v_0^2 + 2ax_A$, with $v_f = 0$. Thus, $x_A = -v_0^2/2a$, or $x_A = -(21 \text{ m/s})^2/(-6 \text{ m/s}^2) = 73.5$ m.

**(A)** To find the maximum $\Delta t$, we note that $x_{max} = 110$ m and $x_{max} = (21 \text{ m/s})\,\Delta t_{max} + x_A$, or 110 m = (21 m/s) $\Delta t_{max}$ + 73.5 m, and $\Delta t_{max} = 1.74$ s. The distance moved before braking started is, of course, just $v_0\,\Delta t = 36.5$ m.

**(B)** If $t = 1.4$ s, then $x = (21 \text{ m/s})(1.4 \text{ s}) + 73.5 \text{ m} = 102.9$ m. The distance to the car is just 110 m − 102.9 m = 7.1 m. To find the time, we need to know the time $t$ during which the truck accelerates. We have $v_f = v_0 + at$, with again $v_f = 0$ and $a = -3$ m/s$^2$. Then $0 = 21 \text{ m/s} - (3 \text{ m/s}^2)t$ and $t = 7$ s. The total time is $t + t = 7 + 1.4 = 8.4$s.

**38.** The car starts with initial velocity zero and acceleration $a_c = 1.4$ m/s$^2$, while the bus has constant velocity $v_b = 12$ m/s.

**(A)** Both travel the same distance $x$ in time $t$, so set $a_c t^2/2 = v_b t$ to give $t = 17$ s.

**(B)** The final velocity of the car is $v = a_c t = 24$ m/s.

**(C)** The average velocity of the car (or the bus's fixed velocity) times 17 s yields $x = 204$ m.

**39. (A)** The time for the coconut to fall 20 m is given by $20 = gt^2/2$, or 2.02 s. Distance $x = (1.5 \text{ m/s})(2.02 \text{ s}) = 3.03$ m.

**(B)** Since you are moving at a fixed speed, the monkey should have dropped the coconut 2.02 s earlier.

**40.** The time to fall from the greater height $h_1$ is $t = 5$ s; the time from the lesser height $h_2$ is $t - 2 = 3$ s. Using $y = at^2/2$, we find

**(A)** the height difference $h_1 - h_2 = g\,(5^2/2 - 3^2/2) = 78$ m

**(B)** $h_1 = 9.8(25)/2 = 123$ m.

**41.** The two boys meet at the same place and time. The time for the slower one to travel a distance $x$ is $x/5$, while the other boy has $t = (100 - x)/7$. Equating the times yields $x = 41.7$ m.

## Chapter 3: Newton's Laws of Motion

**42. (A)** 300 g

**(B)** $w = (300 \text{ g})(980 \text{ cm/s}^2) = 2.94\times10^5$ dyn = 2.94 N.

**(C)** $m = w/g = (20 \text{ N})/(9.8 \text{ m/s}^2) = 2.04$ kg; mass is the same anywhere.

**(D)** $m = w/g = 5 \text{ N}/10 \text{ N/kg} = 0.5$ kg.

**43.** Here we start with the kinematical equation that allows us to find the acceleration $a_x$: $v_x^2 = v_{0x}^2 + 2a_x x$, where $v_x = 0$ when $x = 30$ m and $v_{0x} = 20$ m/s. Solving, we obtain $a_x = -6.67$ m/s$^2$. Finally, we solve for the retarding force, using $F_x = ma_x = (900 \text{ kg}) \times (-6.67 \text{ m/s}^2) = -6000$ N.

**44.** The crate is acted on by two vertical forces—the tension in the rope, $T$, upward and the weight of the crate, $w = mg$, downward. Noting that $w = (20 \text{ kg})(9.8 \text{ m/s}^2) = 196$ N and using $T - w = ma_y$, we get:

**(A)** $a_y = 2.7$ m/s$^2$

**(B)** $a_y = -2.3$ m/s$^2$

**(C)** $a_y = -9.8\ \text{m/s}^2$
**(D)** $a_y = 0$. Note that negative acceleration is downward for our case.

**45.** Letting the $x$ axis be along the direction of motion, we have for the magnitude of the resultant force $F_x = ma_x$. To find $a_x$ we use the kinematical relationship $v_x = v_{0x} + a_x t$, with $v_{0x} =$ 5.0 m/s, $v_x = 2.0$ m/s, $t = 6.0$ s. Solving, we get $a_x = -0.50\ \text{m/s}^2$. Then $F_x = (40\ \text{kg})(-0.50\ \text{m/s}^2) =$ $-20$ N. Noting that the resultant force in the $y$ direction is zero since $a_y = 0$, we have our answer: **F** is 20 N in the direction opposite to the velocity.

**46.** From $F = ma$, with $F = 20$ N and $a = 8.0\ \text{m/s}^2$, we get $m = 2.50$ kg. From $F = m'a'$ and $a' = 24.0\ \text{m/s}^2$, we get $m' = 0.83$ kg. Combining the two masses yields $M = m + m' = 3.33$ kg and $F = MA$ yields A = 6.0 m/s$^2$.

**47.** For the boy to be in equilibrium, the floor must push up on the boy's feet with a force $F$ equal and opposite to the combined weight of the flour and the boy. Let $m$ equal the mass of the boy and $w$ the weight of the flour:

$$F = mg + w = 75(9.8) + 40 = 735 + 40 = 775\ \text{N}$$

**48.** Newton's third law says that the VW owner is right. Each car must pull on the chain with the same magnitude of force, one being the action and the other the reaction. The motion of the VW is a consequence of *all* the forces acting on it, not just the force of the chain. These other forces include, in particular, the frictional force between tires and road, which is quite different for the VW and the Cadillac.

**49.** To obtain the tension $T$ in the string, we apply the second law to the $m = 3.0$ kg package: $T - w = ma_y$, with $w = mg = 29.4$ N. The acceleration $a$ of the package is the same as that of the elevator; it is obtained from the displacement formula $y = v_{0y}t + \frac{1}{2}a_y t^2$, where $v_{0y} = 0$ and $y = 2.0$ m, when $t = 0.60$ s. Solving, we get $a_y = 11\ \text{m/s}^2$. Substituting into our equation for $T$, we get $T = 63$ N.

**50.** The maximum upward force exerted by the scale on the boy is 400 N. The net force on the boy is (400 – 300) N, and this equals $ma$. Using $m = 300/9.8$ yields $a = 3.3\ \text{m/s}^2$.

**51.** When the book of mass $m$ is about to slide, the friction $f = \mu mg$. Friction is the only horizontal force acting, thus $f = ma$. Inserting $\mu = 0.45$ yields $a = mg = 4.4\ \text{m/s}^2$.

**52.** Again choosing upward as positive and letting $N$ represent the force of the scale on the man, we have $N - w = ma_y$. Noting that $w = 700$ N, and that $m = w/g = 71.4$ kg, we solve for $N$ using the values of $a_y$ given:
**(A)** $N = 829$ N
**(B)** $N = 571$ N
**(C)** $N = 0$

**53.** Let the pulling force of the rope be $T$. Using $\Sigma F_x = ma_x$, we have for our case $T\cos 30° - 30\ \text{N} = ma_x$, where $m = 20$ kg.
**(A)** For $a_x = 0$, $T = 35$ N
**(B)** For $a_x = 0.40\ \text{m/s}^2$, $T = 44$ N

**54.** We must find $f$ by use of $F = ma$. But first we must find $a$ from a motion question. We know that $v_0 = 0$, $v_f = 2$ m/s, $t = 4$ s. Using $v_f = v_0 + at$ gives

$$a = \frac{v_f - v_0}{t} = \frac{2 \text{ m/s}}{4 \text{ s}} = 0.50 \text{ m/s}^2$$

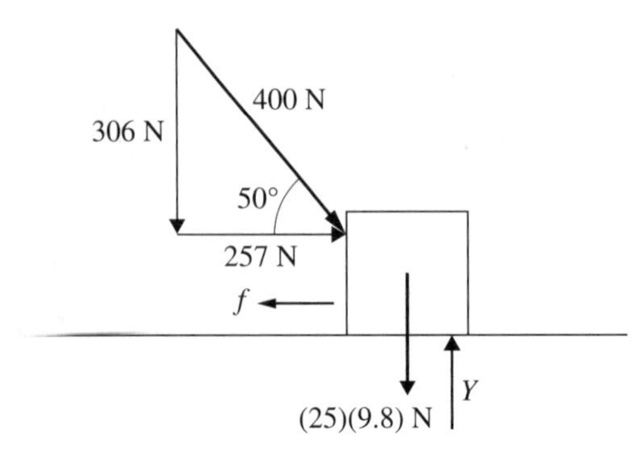


**Figure A3.1**

Now we can write $\Sigma F_x = ma_x$, where $a_x = a = 0.50$ m/s$^2$. From Figure 3.1, this equation is 257 N $- f =$ (25 kg)(0.50 m/s$^2$), or $f = 265$ N. We now wish to use $\mu = f/Y$. To find $Y$, we write $\Sigma F_y = ma_y = 0$, since no vertical motion occurs. From Figure 4.6, $Y -$ 306 N $-$ (25)(9.8) N $= 0$, $Y = 551$ N.

$$\mu = \frac{f}{Y} = \frac{265}{551} = 0.44$$

**55.** In Figure A3.2, we show the three forces acting on the block: the frictional force $f = 60$ N; the normal force $N$, which is perpendicular to the incline; and the weight of the block, $w = mg =$ (12 kg)(9.8 m/s$^2$) = 118 N. We choose the $x$ axis along the incline, with downward as positive.

**(A)** Using $\Sigma F_x = ma_x$, we have $w \sin 40° - f = ma_x$, or (118 N)(0.642) – (60 N) = (12 kg)$a_x$. Solving, we have $a_x = 1.31$ m/s$^2$.

**(B)** To find the time to reach the bottom of the incline, starting from rest, we use $x = v_{0x}t + \frac{1}{2} a_x t^2$, with $v_{0x} = 0$ and $x = 5.0$ m. Solving, we get $t = (7.63 \text{ s}^2)^{1/2} = 2.76$ s.

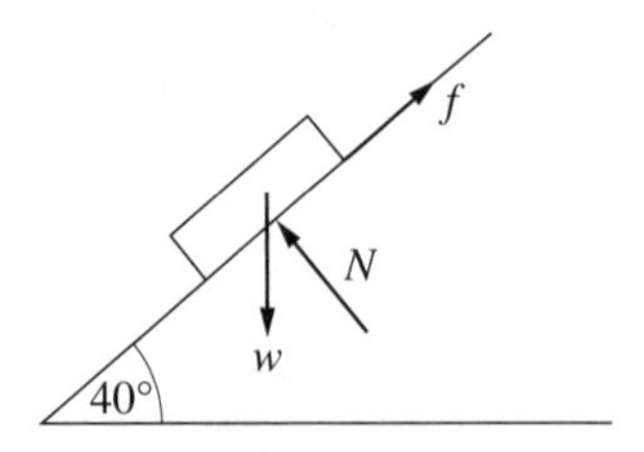


**Figure A3.2**

**56.** Here we assume that the $x$ axis is along the incline and positive upward. If $P$ is the constant force referred to, then from $\Sigma F_x = ma_x$, we have $P - w \sin 30° = ma_x$, *with* $m = 15$ kg and $w = mg = 147$ N.

**(A)** For $a_x = 1.2$ m/s$^2$, $P = 91.5$ N

**(B)** $a_x = -1.2$ m/s$^2$, $P = 55.5$ N

**57.** For the block $\Sigma F_y = 0$ yields $F_N = W = mg$ and $\Sigma F_X = ma$ yields $-mF_N = ma$; hence, $\mu = -a/g$. The uniform acceleration, from $v^2 - v_0^2 = 2ax$, is $a = -1.2^2/2(0.7) = -1.03$ m/s$^2$; then $\mu = 1.03/9.8 = 0.105$.

**58. (A)** The component of the weight down the incline = $mg \sin 30° = 5(9.8)(0.5) = 24.5$ N, while the external force up the plane is $F$. So $F_{net} = ma$ becomes $F - 24.5 = 5(0.20)$, from which $F = 26$ N.

**(B)** A friction force = $\mu F_N$ must be added to 24.5 N down the incline. $F_N = mg \cos 30° =$ 42 N and $\mu F_N = 12.7$ N, so $F$ is larger by this amount; thus, $F = 38$ N.

**59.** The component of the weight down the incline = 8(9.8)(0.5) = 39.2 N. Now $F = ma$ leads to $39.2 - f = 8(0.3)$, so that $f = 36.8$ N. The normal force equals the component of $W$ perpendicular to the incline, 8(9.8)(0.867) = 67.9 N. Therefore, $\mu =$ 36.8/67.9 = 0.54.

**60.** If the block is not to fall, the friction force, $f$, must balance the block's weight: $f = mg$. But the horizontal motion of the block is given by $N = ma$. Therefore,

$$\frac{f}{N} = \frac{g}{a} \quad \text{or} \quad a = \frac{g}{f/N}$$

Since the maximum value of $f/N$ is $\mu_s$, we must have $a \geq g/\mu_s$ if the block is not to fall.

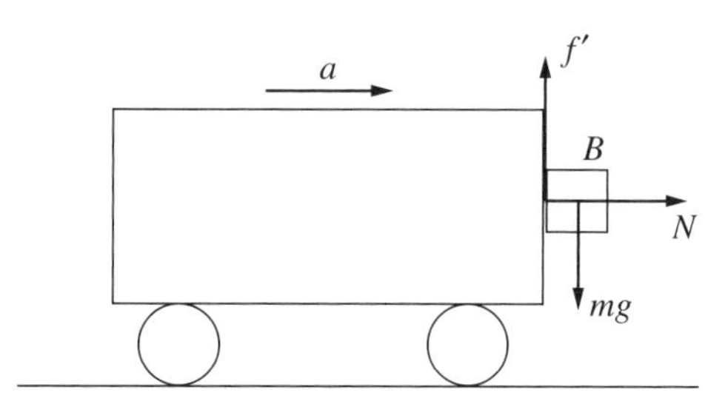


**Figure A3.3**

**61.** First apply Newton's second law to the system as a whole to find the force $F_1$ accelerating both masses upward.

$$F_1 = (m_A + m_B)a = (15 + 11)3 \qquad F_1 = 78 \text{ N}$$

Since $F_1$ is the resultant force, $F_1 = T_1 - m_A g - m_B g$, and the tension $T_1$ is the sum of the weights of $A$ and $B$ plus $F_1$.

$$T_1 = m_A g + m_B g + F_1 = 15(9.8) + 11(9.8) + 78 = 147 + 107.8 + 78 \quad T_1 = 330 \text{ N}$$

Similarly, for mass $B$ only,

$$F_2 = m_B a = 11(3) = 33 \text{ N} \qquad T_2 = m_B g + F_2 = 11(9.8) + 33 = 107.8 + 33 = 140 \text{ N}$$

To check, for block $A$ only,

$$T_1 = m_A g + m_A a + T_2 = 147 + 45 + 140.8 = 330 \text{ N}$$

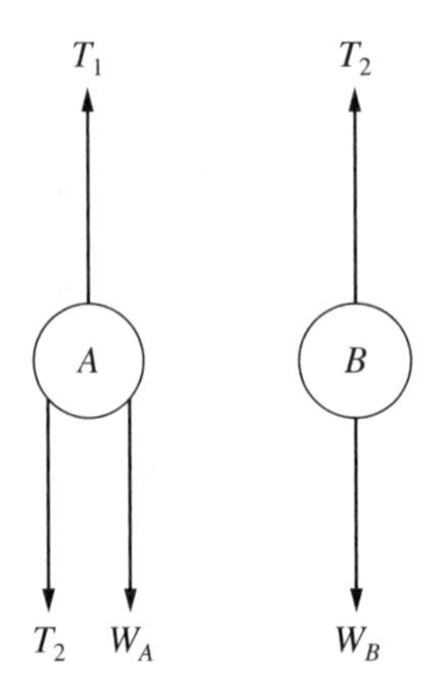


**Figure A3.4**

**62.** Write $F = ma$ for each block, using $f$ as the friction force on each block. Then we obtain $F - f - T = m_2 a$ and $T - f = m_1 a$. Use the given values and solve to find $T = 10$ N and $f = 8.5$ N.

**63.**

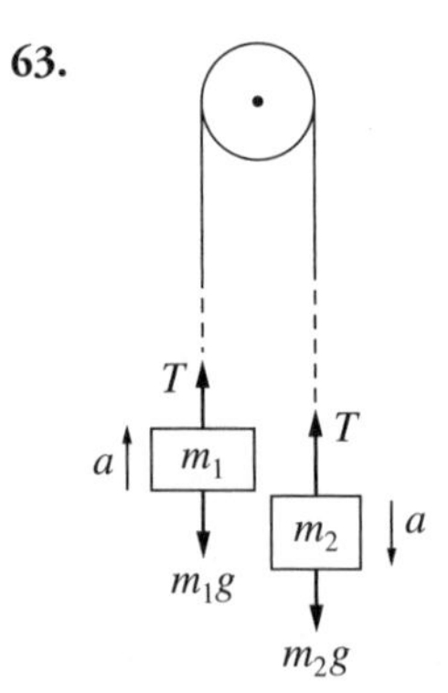


**Figure A3.5**

**(A)** Isolate the forces on each mass and write Newton's second law, choosing *up* as positive: $T - m_1 g = m_1 a$ and $T - m_2 g = -m_2 a$. Eliminating $T$ gives $a = (m_2 - m_1)g/(m_1 + m_2)$. Now use $y - at^2/2$ to find the distance fallen in time $y = 0.63$ m.

**(B)** From the above equations, $T = 2m_1 m_2 g/(m_1 + m_2) = 27$ N.

Mass $B$ will rise and mass $A$ will fall. You can see this by noting that the forces acting on pulley $P_2$ are $2T_2$ up and $T_1$ down. Therefore, $T_1 = 2T_2$ (the inertia-less object transmits the tension). Twice as large a force is pulling upward on $B$ as on $A$.

Let $a$ = downward acceleration of $A$. Then $\frac{1}{2}a$ = upward acceleration of $B$. (As the cord between $P_1$ and $A$ lengthens by 1 unit, the segments on either side of $P_2$ each shorten by $\frac{1}{2}$ unit. Hence, $\frac{1}{2} = s_B/s_A = (\frac{1}{2}a_Bt^2)/(\frac{1}{2}a_At^2) = a_B/a_A$.) Write $\Sigma F_y = ma_y$ for each mass in turn, taking the direction of motion as positive in each case. We have

$$T_1 - 300\text{ N} = m_B(\tfrac{1}{2}a) \qquad \text{and} \qquad 200\text{ N} - T_2 = m_Aa$$

But $m = w/g$, and so $m_A = (200/9.8)$ kg and $m_B = (300/9.8)$ kg. Further, $T_1 = 2T_2$. Substitution of these values in the two equations allows us to compute $T_2$ and then $T_1$ and $a$. The results are

$$T_1 = 327\text{ N} \qquad T_2 = 164\text{ N} \qquad a = 1.78\text{ m/s}^2$$

**64.**

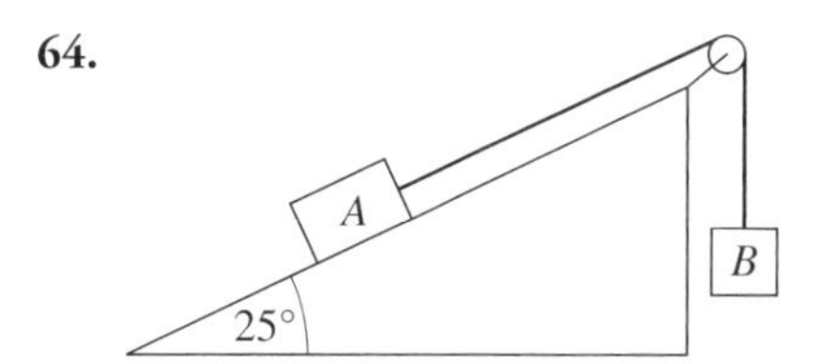


**Figure A3.6**

The situation is as shown in Figure A3.6. We apply Newton's second law to each block separately. For block $B$ we choose downward as positive, while for block $A$ we choose our $x$ axis along the incline with the positive sense upward. This choice allows us to use the same symbol, $a$, for the acceleration of each block. Then for block $B$, $w_b - T = m_ba$, where $m_b$ = 20 kg and $w_b$ = 196 N and $T$ is the tension in the cord. Since the pulley is frictionless, the same tension $T$ will exist on both sides of the pulley. Then for block $A$, $T - w_a \sin 25° = m_aa$, where $m_a$ = 30 kg and $w_a$ = 294 N. We can eliminate the tension $T$ by adding the two equations, which yields $w_b - w_a \sin 25° = (m_a + m_b)a$. Substituting in the known values, we solve, getting $a$ = 1.44 m/s$^2$. The equation for fall from rest is $y = v_{0y}t + \frac{1}{2}a_yt^2$, with $v_{0y} = 0$. Substituting in $a_y$ = 1.4 m/s$^2$ and $t$ = 2 s, we get $y$ = 2.9 m.

**65.** **(A)** Let the tension in the cord on the left be $T_1$ and on the right be $T_2$. The pulleys are assumed to be frictionless. We apply Newton's second law to each of the three blocks, choosing the positive sense of the axis for each block consistently. Thus we choose downward as positive for block $C$, to the right as positive for block $B$, and upward as positive for block $A$. The frictional force on block $B$ is to the left and can be obtained from $f = \mu_kN$, where $\mu_k = 0.20$ and the normal force $N$ equals the weight $w_b$ = 98 N from vertical equilibrium. Thus, $f$ = 19.6 N. For our three equations we have

$$w_c - T_2 = m_ca$$

where $a$ is the acceleration, $m_c$ = 9 kg, and $w_c$ = 88.2 N.

$$T_2 - T_1 - f = m_ba$$
$$T_1 - w_a = m_aa$$

where $m_a$ = 6 kg and $w_a$ = 58.8 N.

As with earlier questions involving cords connecting blocks, the tensions in adjacent equations appear with opposite signs. Adding the three equations eliminates the tensions completely: $w_c - f - w_a = (m_a + m_b + m_c)a$. Note that this is equivalent to a one-dimensional question involving a single block of mass $m_a + m_b + m_c$ acted on by a force $w_c$ to the right and forces $f$ and $w_a$ to the left. Substituting the known masses, weights, and $f$ gives $a = 0.39\ \text{m/s}^2$.

**(B)** Substitute $a$ back into the equations of motion for each block, most conveniently the first and third, to obtain $T_1 = 61$ N, $T_2 = 85$ N.

**66.** Note, $T_1 > T_2$. Consider the free body made up of both masses and the massless cord between them: $T_1 - 1.0\ g = 1.0\ a$; for $T_1 = 15.0$ N, $a = 5.2\ \text{m/s}^2$; for $T_1 = 7.0$ N, $a = -2.8\ \text{m/s}^2$. (The system must be accelerating downward, since $T_1$ could not support the 9.8-N weight if it were in equilibrium.)

**67. (A)** This type of question, as seen in Question 65, can be treated as if it were in one dimension. Thus, Newton's second law takes the form

$$F = ma \quad m_2 g = (m_1 + m_2)a \quad 3(9.8) = (6+3)a \quad 29.4 = 9a \quad a = 3.3\ \text{m/s}^2$$

**(B)** Applying Newton's second law to mass $m_1$ alone,

$$T = m_1 a = 6(3.27) = 20\ \text{N}$$

**68. (A)** See Figure A3.7(*a*). Considering the blocks to move as a unit, $M = m_a + m_b = 8$ kg, $F = Ma = 6$ N $a = 0.75\ \text{m/s}^2$.

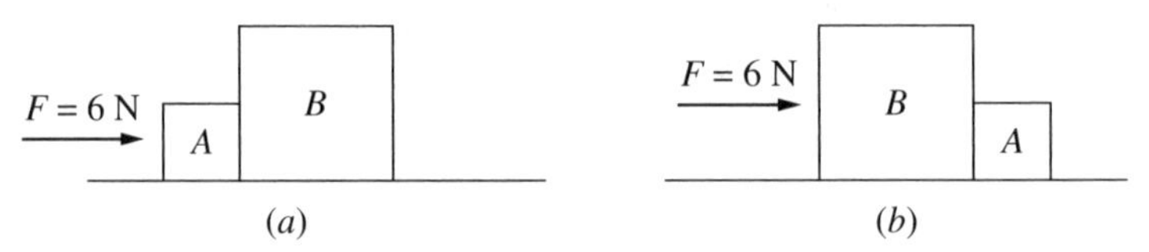


**Figure A3.7**

**(B)** If we now consider block $B$ to be our system, the only force acting on it is the force due to block $F_{ab}$. Then, since the acceleration is the same as in part (*a*), $F_{ab} = M_b a = 4.5$ N. However, we must consider also the case in which we reverse blocks $A$ and $B$, as in Figure A3.7(*b*). As before, considering the blocks as a unit, we have $a = 0.75\ \text{m/s}^2$. Now, however, if we consider block $B$ as our system, we have two forces acting, the force $F$ to the right and the force $F_{ab}$ to the left. Then $F - F_{ab} = M_b a$ and solving we get $F_{ab} = 1.5$ N in magnitude and points to the left. (The answer would be the same if we considered block $A$ to be the system.)

## Chapter 4: Motion in a Plane

**69. (A)** If the water were standing still, the boat's speed past the tree would be 8 km/h. But the stream is carrying it in the opposite direction at 3 km/h. Therefore, the boat's speed relative to the tree is 8 – 3 = 5 km/h.

**(B)** In this case, the stream is carrying the boat in the same direction the boat is trying to move. Hence, its speed past the tree is 8 + 3 = 11 km/h.

**70.** The plane's resultant velocity is the vector sum of two velocities, 500 km/h eastward and 90 km/h southward. These component velocities are shown in Figure A4.1. The plane's resultant velocity is found by use of

$$R = \sqrt{(500)^2 + (90)^2} = 508 \text{ km/h}$$

The angle a $\alpha$ is given by

$$\tan \alpha = \frac{90}{500} = 0.180$$

from which $a$ = 10.2°. The plane's velocity relative to the ground is 508 km/h at 10.2° south of east.

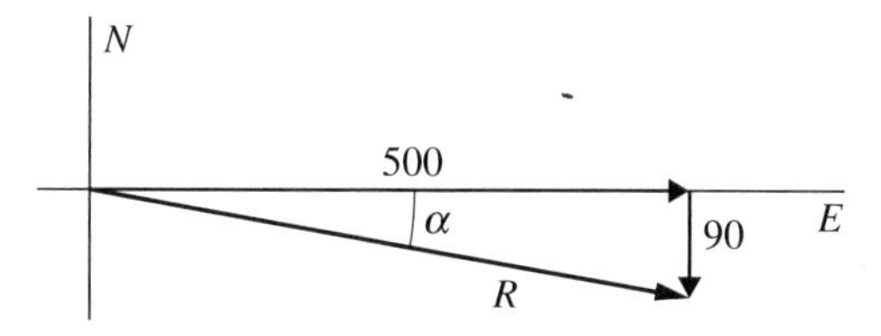


**Figure A4.1**

**71.** Choose downward as positive with origin at edge of table top.

$$v_{0x} = v_0 = 20 \text{ cm/s} \qquad v_{0y} = 0 \qquad a_y = +g = +980 \text{ cm/s}^2 \qquad a_x = 0$$

To find time of fall, $y = v_{0y}t + \frac{1}{2}gt^2$, or 80 cm = 0 + (490 cm/s²)$t^2$; $t$ = 0.40 s. The horizontal distance is gotten from $x = v_{0x}t$ = (20 cm/s)(0.40 s) = 8.0 cm.

**72.** For the electron, the horizontal problem yields the time to hit the screen as $t = x/v_x$ = $0.40/(5 \times 10^7) = 8 \times 10^{-9}$ s. Then in the vertical problem, $y = v_{0y}t + at^2/2$, so $y = 0 + 4.9 \times (64 \times 10^{-18}) = 3.1 \times 10^{-16}$ m. For a droplet, $t$ = 0.40/2 = 0.20 s, and so $y$ = 0.196 m.

**73.**

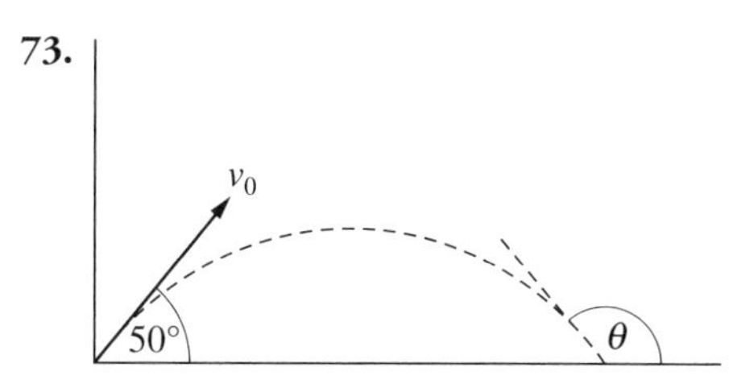


**Figure A4.2**

Choose upward as positive, and place the origin at the launch point (Figure A4.2).

$$v_{0x} = v_0 \cos 50° = (40 \text{ m/s})(0.642) = 25.7 \text{ m/s} \qquad v_{0y} = v_0 \sin 50° = (40 \text{ m/s})(0.766) = 30.6 \text{ m/s}$$

$$a_y = -g = -9.8 \text{ m/s}^2 \qquad a_x = 0$$

To find the time in air, we have $y = v_{0y}t - \frac{1}{2}gt^2$ and since $y = 0$ at the end of flight, $0 = (30.6 \text{ m/s})t - (4.9 \text{ m/s}^2)t^2$, or $4.9t^2 = 30.6t$. The first solution $t = 0$ corresponds to the starting point, $y = 0$. The second solution is not zero and is obtained by dividing out by $t$. $4.9t = 30.6$ and $t = 6.24$ s.

**74.** The horizontal acceleration of a projectile is zero. So we have $x = v_{0x}t = (25.7 \text{ m/s}) \times (6.24 \text{ s}) = 160$ m. The velocity upon hitting the ground has component 25.7 m/s horizontally. Because the object hits the ground at the same height from which it was projected, symmetry dictates that the vertical component of the final velocity is the same as the vertical component of the initial velocity, but downward: $v_{fy} = 30.6$ m/s. To find the angle of the final velocity vector with the horizontal, $\tan\theta = v_{fy}/v_{fx} = (30.6 \text{ m/s})/(25.7 \text{ m/s}) = 50.0°$.

**75.**

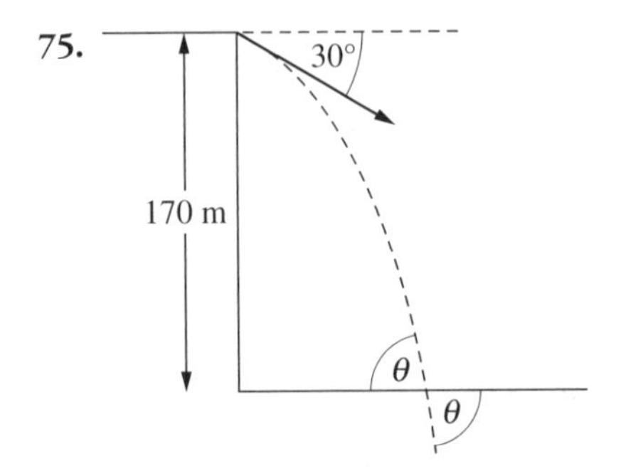


**Figure A4.3**

Choose *downward* as positive and origin at the top edge of building (Figure A4.3).

$$v_{0x} = v_0 \cos 30° = (40 \text{ m/s})(0.866) = 34.6 \text{ m/s}$$
$$v_{0y} = v_0 \sin 30° = (40 \text{ m/s})(0.500) = 20.0 \text{ m/s}$$
$$a_y = g = 9.8 \text{ m/s}^2 \quad a_x = 0$$

We can solve for the time in different ways. Method 1:

$$y = v_{0y}t + \tfrac{1}{2}gt^2 \qquad \text{or} \qquad 170 \text{ m} = (20.0 \text{ m/s})t + (4.9 \text{ m/s}^2)t^2$$

We can solve the quadratic to yield

$$t = \frac{-20 \pm (400 + 3332)^{1/2}}{9.8} = 4.2 \text{ s}$$

(We keep only the positive solution; the negative time corresponds to a time before $t = 0$ when it would have been at ground level if it were a projectile launched so as to reach the starting position and velocity at $t = 0$.)

Method 2: We avoid the quadratic. First find $v_y$ just before impact:

$$v_y^2 = v_{0y}^2 + 2gy \quad \text{or} \quad v_y^2 = (20.0 \text{ m/s})^2 + 2(9.8 \text{ m/s}^2)(170 \text{ m}) \quad v_y = \pm 61 \text{ m/s}$$

For our case, $v_y = 61$ m/s. Next we find $t$:

$$v_y = v_{0y} + gt \quad \text{or} \quad 61 \text{ m/s} = 20.0 \text{ m/s} + (9.8 \text{ m/s}^2)t \quad \text{or} \quad t = 4.2 \text{ s}$$

**76.** $x = v_{0x}t = (34.6 \text{ m/s})(4.2 \text{ s}) = 145$ m. We need the angle that the vector velocity makes with the $x$ axis just before hitting the ground. We avoid having to directly deduce the quadrant this angle is in by solving for the acute angle $\theta$ made with the $x$ axis (positive or negative; above or below).

$$\tan\theta = \left|\frac{v_y}{v_x}\right| = 1.76 \quad \text{and} \quad \theta = 60°$$

Since $v_y$ is negative and $v_x$ is positive, this is clearly the angle below the positive $x$ axis (see Figure 5.2) and equals the angle we are looking for.

**77.**

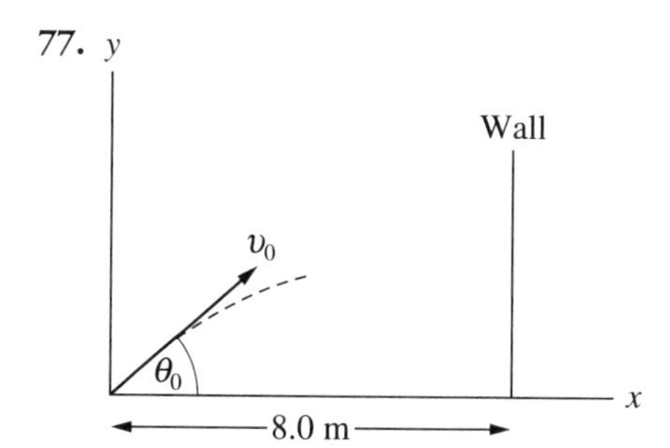


**Figure A4.4**

Setting coordinates as shown in Figure A4.4, with $v_0 = 20$ m/s and $\theta_0 = 40°$, we get

$$v_{0x} = v_0 \cos\theta_0 = (20 \text{ m/s}) \cos 40° = 15.3 \text{ m/s}$$
$$v_{0y} = v_0 \sin\theta_0 = (20 \text{ m/s}) \sin 40° = 13 \text{ m/s}$$

$x = v_{0x}\, t$, and setting $x = 8$ m, we find the time to hit the wall: 8 m = (15.3 m/s)$t$, yielding $t =$ 0.52 s. To find the height at which it hits the wall, we use $y = v_{0y}\, t - \frac{1}{2} gt^2$, with $t = 0.52$ s. This yields $y = (12.8 \text{ m/s})(0.52 \text{ s}) - (4.9 \text{ m/s}^2)(0.52 \text{ s})^2 = 5.33$ m.

**78.**

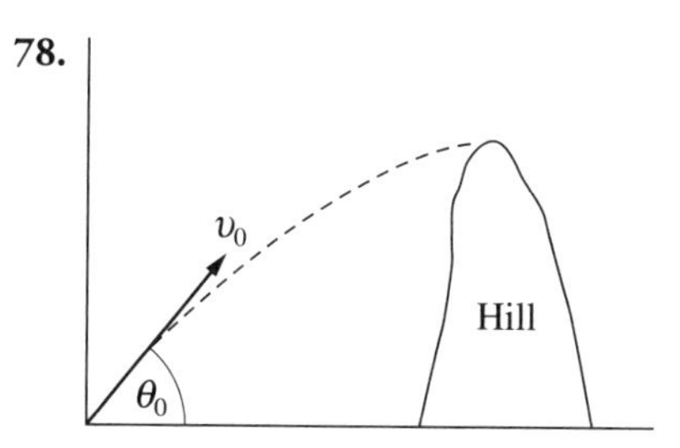


**Figure A4.5**

The situation is as shown in Figure A4.5, $v_0 = 95$ m/s; $\theta_0 = 50°$. At any time, $t, y = v_{0y}t - \frac{1}{2}gt^2$, where $v_{0y} = v_0 \sin\theta_0 = 72.8$ m/s. For $t = 5$ s, we get $y = (72.8 \text{ m/s})(5.0 \text{ s}) - \frac{1}{2}(9.8 \text{ m/s}^2)\times(5.0 \text{ s})^2 = 241$ m. The horizontal distance, $x$, is given by $x = v_{0x}t = v_{0x}t = v_0 \cos\theta_{0t} = (61.1 \text{ m/s})t$. At $t = 5$ s, we have $x = 305$ m.

**79.** Use $y = v_{0y}t + \frac{1}{2}a_y t^2$ to get $100 \text{ m} = 0 + \frac{1}{2}(9.8 \text{ m/s}^2)t^2$, or $t = 4.5$ s. Now use $x = v_x t = (15 \text{ m/s})(4.5 \text{ s}) = 68$ m, since the sack's initial velocity is that of the plane.

**80.** To move horizontally with the cart, the projectile must be fired vertically with a flight time $= x/v_x = (80 \text{ m})/(30 \text{ m/s}) = 2.67$ s. The initial velocity $v_0$ must satisfy $y = v_0 t + at^2/2$, with $y = 0$ and $t = 2.67$ s; thus, $4.9t = v_0$, and so $v_0 = 13.1$ m/s at $\theta = 90°$.

**81.** **(A)** At the point of impact, $y = -35$ m and $x = R$. From $y = -35 = (80 \sin 25°)t = \frac{1}{2}(9.8)t^2$, $t = 7.8$ s. Then $x = R = (80 \cos 25°)(7.814) = 570$ m.

**(B)** At impact, $v_y = 80 \sin 25° - (9.8)(7.814) = -42.77$ m/s and $v_x = v_{0x} = 80 \cos 25° = 72.5$ m/s. Thus, $v = (42.77^2 + 72.5^2)^{1/2} = 84$ m/s and $\tan\beta = -42.77/72.5$, or $\beta = 31°$ below horizontal.

**82.**

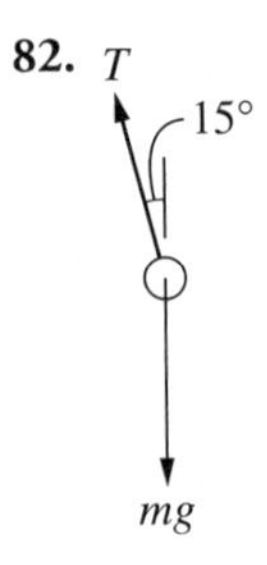


**Figure A4.6**

$$T\cos 15° = mg \quad T\sin 15° = \frac{mv^2}{r} \quad \text{and hence,} \quad \tan 15° = \frac{v^2}{rg}$$

Since $r = 24 \sin 15° = 24(0.259) = 6.22$ cm,

$$\tan 15° = \frac{v^2}{6.22(980)} \qquad v = 40 \text{ cm/s}$$

**83.** First draw a diagram showing the forces (Figure A4.7). If $mg$ is the weight of the automobile, then the normal force is $N = mg$. The frictional force supplies the centripetal force $F_c$.

$$F_c = \mu_s N = 0.81\, mg$$

$$\text{Also, } F_c = \frac{mv^2}{r} \qquad 0.81\, mg = \frac{mv^2}{80} \qquad v^2 = 0.81 \times 80 \times 9.8 = 25 \text{ m/s}$$

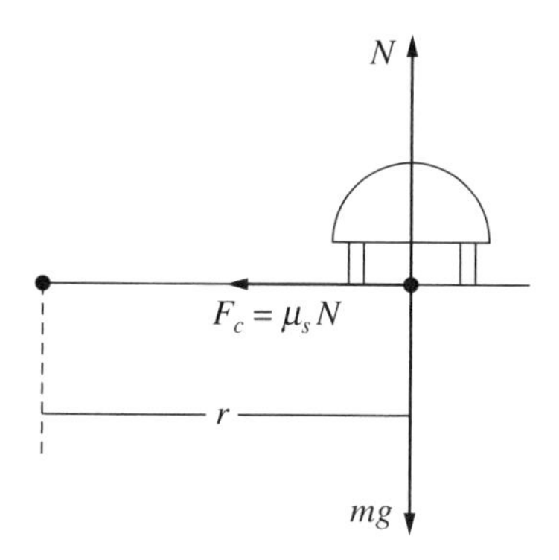


**Figure A4.7**

**84.** **(A)** $\dfrac{v^2}{R_m} = \dfrac{GM_m}{R_m^2} = g_m \quad v = \sqrt{g_m R_m} = \sqrt{(1.63\ \text{m/s}^2)(1.74\times10^6\ \text{m})}$

$= 1.68\times10^3\ \text{m/s} = 1.68\ \text{km/s}$

**(B)** The circumference of the orbit is

$$d = 2\pi R_m = (6.28)(1.74\times10^6\,\text{m}) = 1.09\times10^4\,\text{km}$$

so the period is

$$t = \frac{d}{v} = \frac{1.09\times10^4\ \text{km}}{1.68\ \text{km/s}} = 6.5\times10^3\,\text{s} = 108\ \text{min}$$

**85.** $T\sin\theta = \dfrac{mv^2}{T}$; $T\cos\theta = mg$ where $T$ = tension.

Thus, $\tan\theta = \dfrac{v^2}{rg}$ or $v = \sqrt{rg\tan\theta} = 23\ \text{m/s} = 82.5\ \text{km/h}$

**86.** First we change miles per hour to feet per second:

$$350\ \text{mi/h} = 513\ \text{ft/s}$$

Substitute values:

$$F = \frac{180\ \text{lb}\times(513\ \text{ft/s})^2}{32.2\ \text{ft/s}^2\times2000\ \text{ft}} + 180\ \text{lb} = 915\ \text{lb}$$

**87.** To experience weightlessness, the normal force of the seat on the ride must equal zero. Thus, the gravitational force $mg$ must equal the required centripetal force $mv^2/r$. Equating the two and solving for $v$ gives 14 m/s.

**88.** The maximum tension will occur at the bottom when the cable must furnish a force $mg + mv^2/r$. To reach the bottom, the mass falls a distance $h = (15 - 15 \cos 37°) = 3.0$ m. By conservation of energy, its speed there will be $v = (2gh)^{1/2} = (6g)^{1/2}$. Therefore, the tension will be $T = 200g + 200(6g)/15 = 2740$ N.

**89.** In American engineering units, 16 lb $\Rightarrow$ 0.497 slug, and Newton's law of gravitation has the form $F = G\,[(m_1 m_2)/d^2]$, with $G = 3.44\times10^{-8}$ lb·ft²/slug². Thus,

$$F = \left(3.44\times10^{-8}\,\frac{\text{lb}\cdot\text{ft}^2}{\text{slug}^2}\right)\left(\frac{0.497\text{ slug}\times0.497\text{ slug}}{(2\text{ ft})^2}\right) = 2.12\times10^{-9}\text{ lb}$$

**90.** **(A)** The mass is the same as that on the earth's surface,

**(B)** As long as we are outside the earth's surface, the weight (force of gravity) varies inversely as the square of the distance from the center of the earth. Indeed $w = GmM/r^2$, where $m$, $M$ are the masses of object and earth, respectively, and $r$ is the distance to the center of the earth. Thus, $w_2/w_1 = r_1^2/r_2^2$, since $G$, $m$, $M$ are constant in this question. For our case we set $r_1 = 6370$ km and $r_2 = 6530$ km and $w_1 = (20\text{ kg})(9.8\text{ m/s}^2) = 196$ N. This gives $w_2 = 187$ N.

**91.** Newton's law of gravitation, $w = GmM/r^2$, gives $w_2/w_1 = (M_2/M_1)(r_1^2/r_2^2)$. Letting 1 refer to earth and 2 refer to Mars, we have $w_2 = 0.11(6370/3440)^2(200\text{ N}) = 75$ N. The acceleration is derived from $w_2/w_1 = g_2/g_1$, or $g_2 = (75/200)(9.8\text{ N}) = 3.7\text{ m/s}^2$.

**92.** The gravitational attraction between the earth and moon provides the centripetal force; therefore, $mv^2/r = GMm/r^2$, where $M$ is the earth's mass. Then $M = v^2r/G = \omega^2r^3/G$. Now $\omega = 1$ rev/27 days $= 2.7\times10^{-6}$ rad/s, $r = 3.8\times10^8$ m, and $G = 6.7\times10^{-11}$ in SI. Solving for $M$, it is $6.0\times10^{24}$ kg.

**93.** Let $m$ denote the moon's mass, $M_s$ the sun's mass, $M_e$ the earth's mass, $r_{ms}$ the center-to-center distance from the sun to the moon, and $r_{me}$ the center-to-center distance from the earth to the moon. We let $F_{ms}$ denote the magnitude of the gravitational force exerted on the moon by the sun, and $F_{me}$ denote the magnitude of the gravitational force exerted on the moon by the earth. Then $F_{ms} = GM_s m/r_{ms}^2$ and $F_{me} = GM_e m/r_{me}^2$, so that

$$\frac{F_{ms}}{F_{me}} = \frac{M_s r_{me}^2}{M_e r_{ms}^2}$$

Using the given numerical values, we find $F_{ms}/F_{me} = 2$.

**94.** The satellite must have the same angular velocity, $\omega = 1$ rev/day $= 7.27\times10^{-5}$ rad/s, about the earth's center as has the earth itself. As the gravitational force is the centripetal force that keeps the satellite in orbit, $GMm_s/(R_e + h)^2 = m_s\omega^2(R_e + h)$. First solve for $(R_e + h)$; then find $h = 35800$ km, or about $5.6/R_e$.

## Chapter 5: Work and Energy

**95.** **(A)** In each case take the component of the force in the direction of the displacement: (85 cos 30° N) (0.70 m) = 51.5 J, (60 cos 45° N)(0.70 m) = 29.7 J.

**(B)** Work is a scalar, so add the work done by each force to give 81.2 J.

**96.** Because horizontal speed is constant, the carton is in horizontal equilibrium: $F = f = \mu F_N$. Normal force is the weight, 20(9.8) = 196 N. Therefore, $W = Fx = 0.60(196) \times (3.0) = 353$ J.

**97.** The work done by the force is $xF \cos 37°$, where $F \cos 37° = f = \mu F_N$. In this case $F_N = mg - F \sin 37°$, so that $F = \mu mg/(\cos 37° + \mu \sin 37°)$. For $\mu = 0.40$ and $m = 20$ kg, $F = 75.4$ N and $W = (75.4 \cos 37°)(8.0) = 482$ J.

**98.** $W = (F \cos 37°)(x) = F_x x$; thus, $F \cos 37° = \mu F_N$, as in Question 97, but now $F_N = mg + F \sin 37°$; solve for $F$: $F = \mu mg/(\cos 37° - \mu \sin 37°) = 140$ N and $F_x = 112$ N. Thus, $W = 112(8.0) = 896$ J.

**99.** To lift a 3-kg object at constant speed, an upward force equal in magnitude to its weight, $mg = (3)(9.8)$ N, must be exerted on the object. The work done by this force is what we refer to as the work done against gravity: work against gravity $= mgh = [(3)(9.8)\text{ N}] \times (0.40\text{ m}) = 12$ J.

**100. (A)** The lifting force is in the direction of the displacement and just balances the weight. $F = mg = 39.2$ N. $W = Fh = (39.2\text{ N})(1.5\text{ m}) = 58.8$ J.

**(B)** If the object is lowered, $F$ is opposite to the displacement; $W = -Fh = -(39.2\text{ N})(1.5\text{ m}) = -58.8$ J.

**101.** The total work is the area under the curve. By counting entire squares and parts of squares, we estimate about 34 squares under the curve, so work is about (34 squares)(40 J/square) = 1360 J.

**102.** $W = \Delta K = 0 - \frac{1}{2}mv^2 = -\frac{1}{2}(1200\text{ kg})(30\text{ m/s})^2 = -540\text{ kJ}$, $W = -fx = -(6\text{ kN})x$; $x = 90$ m.

**103.** Work done by all forces other than gravity equals the combined change in gravitational potential energy and kinetic energy. Since the force $F$ pushing the cart up the incline is the only such force doing work, we have

$$W_F = \Delta U + \Delta K = (mgh - 0) + (0) = (200\text{ kg})(9.8\text{ m/s}^2)(1.5\text{ m}) = 2900\text{ J}$$

**104.** Now we must consider the work done by the frictional force, $f = 150$ N, as well as that done by $F$ (see Question 103). Thus, we have $W_F + W_f = \Delta U + \Delta K = 2.94$ kJ. Noting that $W_f = -(150\text{ N})(7\text{ m}) = -1.05$ kJ, we get $W_F = 3.99$ kJ.

**105.** Between $A$ and $C$, nonconservative work $W' = -fh'$ is done on the rock.

$$\Delta K + \Delta U = W' \quad 0 + [mg(-h') - mgh] = -fh' \quad f = \frac{mg(h+h')}{h'} = \frac{20(16.6)}{0.6} = 553\text{ N}$$

**106.** $\Delta U + \Delta K = 0$; $\Delta U = -mgh = -7.35$ J and $\Delta K = 7.35$ J; $\Delta K = (\frac{1}{2}mv_b^2 - \frac{1}{2}mv_t^2)$, with $v_t = 3$ m/s, so $\frac{1}{2}mv_b^2 = 7.35\text{ J} + 2.25\text{ J} = 9.60$ J. Solving, we get $v_b = 6.20$ m/s.

**107.** **(A)** The tension in the cord does no work. When the ball reaches point *C*, it has lost $mgh_a$ in potential energy, where $h_a = 0.75$ m, and in its place gained $\frac{1}{2}mv_c^2$, in kinetic energy, with $mgh_a = \frac{1}{2}mv_c^2$. The *m*'s cancel, leading to $v_c^2 = 2gh_a = 2(9.8 \text{ m/s}^2)(0.75) = 14.7 \text{ m}^2/\text{s}^2$, or $v_c = 3.83$ m/s.

**(B)** With point *C* as the zero reference for potential energy, the conservation law gives $mgh_a + \frac{1}{2}mv_v^2 = mgh_b + \frac{1}{2}mv_b^2$, with $v_a = 0$, $h_a = 0.75$ m, and $h_b = (0.75 \text{ m})(1 - \cos 37°) = 0.15$ m. Our equation becomes $v_b^2 = 2gh_a - 2gh_b = 11.76 \text{ m}^2/\text{s}^2$, and $v_b = 3.43$ m/s.

**108.** We use $v_A$ and $v_B$ to denote the speeds of the car at points *A* and *B*, and we use $h_A$ and $h_B$ to denote the elevations. If the car has mass *m*, the equation expressing energy conservation is $\frac{1}{2}mv_B^2 + mgh_B = \frac{1}{2}mv_A^2 + mgh_A$.

Since $v_A = 0$, we find $v_B^2 = 2g(h_A - h_B)$. The required centripetal force at *B* is $mv_B^2/R$, where *R* is the radius of curvature of the track at *B*. For the normal force exerted by the track to be positive, the centripetal force must be smaller in magnitude than the car's weight *mg*. That is, we must have

$$\frac{mv_B^2}{R} \leq mg$$

Solving for *R*, we find

$$R \geq R_{\text{min}} = \frac{v_B^2}{g} = 2(h_A - h_B)$$

With $h_A = 15.0$ m and $h_B = 5.0$ m, we obtain $R \geq 2(10) = 20$ m.

**109.** Since the additional 0.20 kg stretched the spring by 0.10 m, $k = F/x = 0.20(9.8)/0.10 = 20$ N/m.

**110.** The spring constant $= k = F/x = 0.20(9.8)/0.10 = 19.6$ N/m, the work is $kx^2/2 = 19.6(0.05)^2/2 = 0.0245$ J for the 5-cm case. The additional work is $[19.6(0.10)^2/2] - 0.0245 = 0.0735$ J.

**111.** $(\text{PE}_b + \text{PE}_s)_{\text{initial}} = (\text{PE}_b + \text{PE}_s)_{\text{final}}$, since KE = 0 at top and bottom. Then $m(9.8)(0.6) + 0 = m(9.8)(0.10) + \frac{1}{2}(2.4 \times 10^3)(0.15)^2$

Solving, $m = 5.5$ kg.

## Chapter 6: Power, Impulse, and Momentum

**112.** $P = \dfrac{W}{t} = \dfrac{Fx}{t} = \dfrac{mgx}{t} = \dfrac{(4.0 \times 10^3)(9.8)(1.0)}{2.0} = 20{,}000 \text{ W} = 20 \text{ kW}$

**113.** Here the power must equal the time rate of increase of gravitational potential energy, since the kinetic energy remains the same. Then $P = mg(\Delta h/\Delta t)$. But $\Delta h/\Delta t = v \sin \theta$, where $v$ is the velocity along the incline and $\theta$ is the angle of the incline. Thus, $P = (1000 \text{ kg}) \times (9.8 \text{ m/s}^2)(20 \text{ m/s})(0.03) = 5.9$ kW. Noting that 0.746 kW = 1 hp, we have $P = 7.9$ hp.

**114.** The work done by the motor each second is

$$\text{power} = mgh + \frac{1}{2}mv^2$$

where $m$ is the mass of grain discharged (and lifted) each second. We are given that $m = 2.0$ kg, $v = 3.0$ m/s, and $h = 12$ m. Substitution gives

$$\text{power} = 244 \text{ W}$$

The motor must have an output of at least 244 W.

**115.** Impulse = change in momentum, which for our case is (3.0 kg)(40 m/s) – (3.0 kg) × (–50 m/s) = 270 kg · m/s (i.e., to the right). Since the only horizontal force on the block is the spring force, the spring impulse is thus 270 N · s (to the right). Noting impulse = force × time and that $t = 0.020$ s, we have average force = 13.5 kN to the right.

**116.** Consider the system (block + bullet). The velocity, and hence the momentum, of the block before impact is zero. The momentum conservation law tells us that

$$\text{momentum of system before impact} = \text{momentum of system after impact}$$

$$(\text{mass}) \times (\text{velocity of bullet}) + 0 = (\text{mass}) \times (\text{velocity of block + bullet})$$

$$(0.008 \text{ kg})v + 0 = (9.008 \text{ kg})(0.40 \text{ m/s})$$

where $v$ is the velocity of the bullet. Solving gives $v = 450$ m/s.

**117.** Apply the law of conservation of momentum to the system consisting of the two masses.

$$\text{momentum before impact} = \text{momentum after impact}$$

$$(0.016 \text{ kg})(0.30 \text{ m/s}) + (0.004 \text{ kg})(-0.50 \text{ m/s}) = (0.020 \text{ kg})v$$

Note that the 4-g mass has negative momentum. Solving gives $v = 0.14$ m/s.

**118. (A)** Apply momentum before = momentum after: $(0.020)(50) = 7.02v$, which gives $v = 0.14$ m/s.

**(B)** Equate $K$ right after the collision to work against friction: $[(7.02)(0.142)^2]/2 = f(1.5)$, which leads to $f = 0.047$ N.

**119.** Choose downward as positive. Let $u$ and $v$ be the speeds of the ball just before and after the collision. Then, from conservation of mechanical energy from the time it is dropped to just before impact, we have $\frac{1}{2}mu^2 = mgh_1$, or $u^2 = 2gh_1 = 78.4 \text{ m}^2/\text{s}^2$, or $u = 8.85$ m/s. Repeating from the moment of rebound to the highest rise point, we have $\frac{1}{2}mv^2 = mgh_2$, or $v^2 = 2gh_2 = 49 \text{ m}^2/\text{s}^2$, or $v = 7.0$ m/s. Then, if $P$ is the momentum imparted to the floor (i.e., to the earth), $mu = -mv + P$, or $P = m(u + v) = 16$ kg · m/s.

**120.** Momentum is conserved, so

$$(0.8\times5)+(1.2\times-4)=(0.8\times-4)+1.2V_f \qquad \text{and} \qquad V_f = 2 \text{ m/s}$$

Coefficient of restitution is

$$e = \left|\frac{\text{velocity of separation}}{\text{velocity of approach}}\right| = \left|\frac{2-(-4)}{5-(-4)}\right| = \frac{6}{9} = 0.7$$

from which 24 m/s $= v_2 - v_1$. Combining this with the momentum equation found above gives $v_2 = -4$ m/s and $v_1 = -28$ m/s. (B) In this case, $v_1 = v_2 = v$ and so the momentum equation becomes $3v = -36$ m/s, or $v = -12$ m/s. (C) Here, $e = 1$ and so

$$e = \frac{v_2 - v_1}{u_1 - u_2} \quad \text{becomes} \quad 1 = \frac{v_2 - v_1}{12 - (-24)}$$

from which $v_2 - v_1 = 36$ m/s. Adding this to the momentum equation gives $v_2 = 0$. Using this value for $v_2$ then gives $v_1 = -36$ m/s.

**121.** Consider first the collision of block and bullet. During the collision, momentum is conserved, so

$$\text{momentum just before} = \text{momentum just after } (0.015 \text{ kg})v + 0 = (3.015 \text{ kg})V$$

where $v$ is the initial speed of the bullet and $V$ is the speed of block and bullet just after collision.

After the collision, mechanical energy is conserved:

$$\text{KE just after collision} = \text{final GPE } \tfrac{1}{2}(3.015 \text{ kg})V^2 = (3.015 \text{ kg})(9.8 \text{ m/s}^2)(0.10 \text{ m})$$

From this we find $V = 1.40$ m/s. Substituting this in the momentum equation gives $v = 281$ m/s for the speed of the bullet.

**122.** Momentum before collision = momentum after collision

$$0.005 \times 250 = (2.495 + 0.005)V \quad V = 0.5 \text{ m/s}$$

Once the collision is completed, we have conservation of mechanical energy.

$$\text{KE at start} = \text{PE in spring at end} \qquad \tfrac{1}{2}(2.5)(0.5)^2 = \tfrac{1}{2}(40)x \qquad x = 125 \text{ mm}$$

**123.** $\Sigma p = 0$ leads to $0.015v = 2.015V$. The velocity $V$ of bullet and block is, using the energy-conservation equation, $2.015V^2/2 = 2.015(9.8)(1.30)$, which leads to $V = 5.05$ m/s. The bullet velocity $v$ is, then, 678 m/s.

**124.** $m_1 U_1 + m_2 U_2 = (m_1 + m_2)\mathbf{v}$. Let $x$ be eastward and $y$ northward. For the $x$ component, (1200 kg)(30.0 m/s) + (3600 kg)(20.0 m/s)(cos 60°) = (4800 kg)$v_x$, or $v_x = 15$ m/s. For the $y$ component, 0 + (3600 kg)(20.0 m/s)(sin 60°) = (4800 kg)$v_y$, or $v_y = 13.0$ m/s; $v = (v_x^2 + v_y^2)^{1/2} = 19.8$ m/s; and $\tan\theta = v_y/v_x$ yields $\theta = 40.9°$ north of east.

**125.** The momentum of the system is conserved during the explosion.

$$\text{momentum before} = \text{momentum after}$$

$$0 = (3.73 \times 10^{-23} \text{ kg})(v) + (6.6 \times 10^{-27} \text{ kg})(1.5 \times 10^7 \text{ m/s})$$

where the mass of the remaining nucleus is $3.73 \times 10^{-25}$ kg and its recoil velocity is $v$. Solving gives

$$v = -\frac{(6.6 \times 10^{-27})(1.5 \times 10^7)}{3.73 \times 10^{-25}} = -2.7 \times 10^5 \text{ m/s}$$

**126. (A)** Because the spring force is an internal force, the center of mass remains at rest.

**(B)** The momentum before and after is zero; thus, $0.2(v_{0.2}) = 0.5(v_{0.5})$. Also, since the table is frictionless, the spring energy $U_s = 3.0\text{ J} = [0.2(v_{0.2})^2]/2 + [0.5(v_{0.5})^2]/2$. Solving between the two expressions gives $v_{0.5} = 1.9$ m/s.

**127.** The forces on the object being considered (the beam) are also shown in Figure 127.1. Because the beam is uniform, its center of gravity is at its geometrical center. Thus, the weight of the beam (200 N) is shown acting at the beam's center. The forces $F_1$ and $F_2$ are exerted on the beam by the supports.

We have two equations to write for this equilibrium situation: $\Sigma F_y = 0$ and $\Sigma\tau = 0$. There are no *x*-directed forces on the beam. $\Sigma F_y = 0$ becomes $F_1 + F_2 - 200\text{ N} - 450\text{ N} = 0$. An axis for computing torques may be chosen arbitrarily. Choose it at *A*, because the unknown force $F_1$ will go through it and exert no torque. The torque equation is, then, $-(200\text{ N})(L/2) - (450\text{ N})(3L/4) + (F_2)(L) = 0$. Dividing through the equation by $L$ and then solving for $F_2$ gives $F_2 = 438$ N.

To find $F_1$, we substitute the value of $F_2$ in the first equation and obtain $F_1 = 212$ N.

**128.** The acting forces are shown in Figure 128.1. We assume that the support point is at a distance $x$ from one end. Let us take the axis point to be at the position of the support. Then the torque equation, $\Sigma\tau = 0$, becomes

$$(200\text{ N})(x) + (100\text{ N})\left(x - \frac{L}{2}\right) - (500\text{ N})(L - x) = 0$$

This simplifies to $(800\text{ N})(x) = (550\text{ N})(L)$, and so we find that $x = 0.69L$. The support should be placed 0.69 of the way from the lighter-loaded end.

To find the load $S$ held by the support, we use $\Sigma F_y = 0$, which gives $S - 200\text{ N} - 100\text{ N} - 500\text{ N} = 0$, from which $S = 800$ N.

**129.** In Figure A6.1, we represent the force exerted by the boy as $P$ and that by the man as $3P$. Take the axis point at the left end. Then the torque equation becomes $-(800\text{ N})(x) - (100\text{ N})(L/2) + (P)(L) = 0$.

A second equation we can write is

$$\sum F_y = 0 \quad \text{or} \quad 3P - 800\text{ N} - 100\text{ N} + P = 0$$

from which $P = 225$ N. Substitution of this value in the torque equation gives $(800\text{ N})(x) = (225\text{ N})(L) - (100\text{ N})(L/2)$, from which $x = 0.22\,L$. The load should be hung 0.22 of the way from the man to the boy.

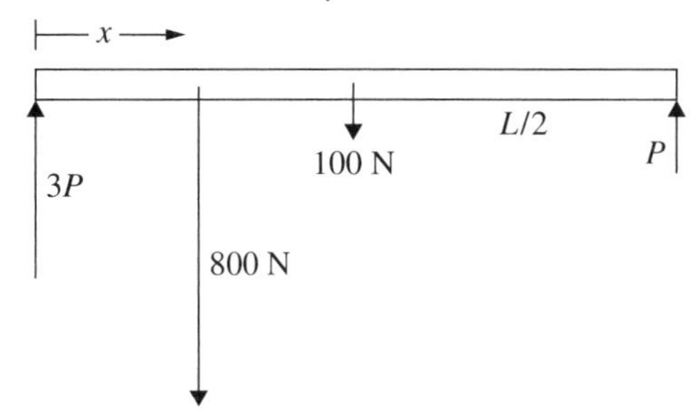


**Figure A6.1**

**130.** For equilibrium, $\Sigma F_y = 0$ and so the equilibrant is of magnitude $P = 400\text{ N} + 200\text{ N} + 300\text{ N} = 900\text{ N}$. Because the board is to be in equilibrium, we are free to choose the axis anywhere. Choose it at end $A$. Then $\Sigma\tau = 0$ gives

$$(P)(x) - (400\text{ N})(3L/4) - (200\text{ N})(L/2) - (300\text{ N})(L/3) = 0$$

Using $P = 900$ N, we find that $x = 0.56L$. The required force is 900 N upward at $0.56L$ from the left end.

**131.** Taking the torques about the attachment point for $W$, one has $-T_1(0.4L) + T_2(0.3L) + 500(0.2L) = 0$ so that $T = 1000$ N, with $T = T_1 = T_2$. From $\Sigma F_y = 0$, one has $2T - W - 500 = 0$, so $W = 1500$ N.

## Chapter 7: Rotational Motion

**132.** **(A)** Use the equation of angular motion, where $\omega_0$ is zero:

$$\omega = \omega_0 + \alpha t \quad 36 = \alpha(6.0) \quad \alpha = 6\text{ rad/s}^2$$

**(B)** Use the equation of angular motion, where $\theta_0$ and $\omega_0$ are zero:

$$\theta = \theta_0 + \omega_0 t + \tfrac{1}{2}\alpha t^2 = \tfrac{1}{2}(6)(6^2) = 108\text{ rad}$$

or use

$$\theta = \theta_0 + \frac{\omega_0 + \omega}{2}t = \frac{0+36}{2}(6) = 108\text{ rad}$$

$$\left(\bar{\omega} = \frac{\omega_0 + \omega}{2} \text{ for uniform angular acceleration}\right)$$

**133.** **(A)** The angular acceleration may be found from $\omega^2 = \omega_0^2 + 2\alpha(\theta - \theta_0)$:

$$\alpha = \frac{\omega^2 - \omega_0^2}{2(\theta - \theta_0)} = \frac{0^2 - [(30)(2\pi)]^2}{2(60)(2\pi)} = -47\text{ rad/s}^2$$

**(B)** The time is found from $\omega = \omega_0 + \alpha t$:

$$t = \frac{0 - (30)(2\pi)}{-47} = 4.0\text{ s}$$

**134.** The respective angular quantities are $\theta = 25/0.3 = 83\text{ rad} = 13.3\text{ rev}$; also $\omega_0 = 2.0\text{ rev/s}$ and $\omega = 0$. Use $\omega^2 - \omega_0^2 = 2\alpha\theta$ to find $\alpha = -0.15\text{ rev/s}^2$, or $\alpha = -0.94\text{ rad/s}^2$.

**135.** First find linear acceleration and distance from $v = v_0 + at$ and $s = \bar{v}t$ to be $a = 0.75\text{ m/s}^2$ and $s = 150$ m. Then transform to angular quantities through $\alpha = a/r = a/0.33\text{ m} = 2.3\text{ rad/s}^2$ and $\theta = s/r = 450\text{ rad} = 72\text{ rev}$.

**136.** **(A)** At $t = 15$ s,

$$\omega = 1200\text{ rev/min} \times 1\text{ min/60 s} \times 2\pi\text{ rad/1 rev} = 40\pi\text{ rad/s}$$

$$\alpha = \frac{\omega - \omega_0}{t} = \frac{(40\pi - 0)\text{ rad/s}}{15\text{ s}} = 8.38\text{ rad/s}^2$$

**(B)** $\tau = I\alpha = 1.6\times10^{-3}\ \text{kg}\cdot\text{m}^2 \times 8.38\ \text{rad/s}^2 = 0.0134\ \text{m}\cdot\text{N}$

**(C)** $\theta = \omega_{\text{avg}} t = \dfrac{(40\pi + 0)\ \text{rad/s}}{2} \times 15\ \text{s} = 942\ \text{rad}$

**(D)** $W = \tau\theta = 0.0134 \times 942 = 12.6\ \text{J}$

or, by the work-energy principle,

$$W = \text{KE} = \tfrac{1}{2} I\omega^2 = \tfrac{1}{2}(1.6\times10^{-3})(40\pi)^2 = 12.6\ \text{J}$$

**137.** Take torques about the pivot point; $30(9.8)(2) - 20(9.8)(2) = I\alpha$. But $I = 20(4) + 30(4) = 200\ \text{kg} \times \text{m}^2$. Substituting gives $\alpha = \frac{196}{200} = 0.98\ \text{rad/s}^2$.

**138.** Torque = (force)(lever arm) = $mgL \sin\theta$, since $r = mr^2\alpha$ with $r = L$ in this case; one has $\alpha = (g \sin\theta)/L$.

**139.** By definition, $I = \Sigma m_i r_i^2 = \Sigma m_i b^2 = 8mb^2$. By comparison with $I = (\Sigma m_i)k^2$, one has $k = b$. Since $\tau = I\alpha$, one finds $\tau = 8mb^2\alpha$. $I_{AA'} = (2m)(b^2) + (2m)(b^2) = 4mb^2$, which is $k_{AA'}^2(8m)$, so one finds that $k_{AA'} = b/2^{1/2}$ and $\tau_{AA'} = 4mb^2\alpha$.

**140.** Only the rods along the $x$ and $y$ axes give rise to a moment of inertia about the $z$ axis. From the parallel-axis theorem, the moment of inertia of a rod about its end is $ML^2/12 + M(L/2)^2 = ML^2/3$. So, since moments of inertia adds, $I_z = 2ML^2/3$.

**141. (A)** First treat each half separately. For the wood, $I$ about the end is, from the parallel-axis theorem, $[m_w(L/2)^2]/12 + m_w(L/4)^2 = (m_w L^2)/12$. For the brass, it would be $(m_b L^2)/12$. Adding these we obtain in the first part $(m_w + m_b)(L^2/12)$.

**(B)** The brass contribution to $I$ about the wood end is $[m_b(L/2)^2]/12 + m_b(3L/4)^2 = (7m_b L^2)/12$, where we used the parallel-axis theorem again. To this we add $(m_w L^2)/12$, so $I = [L^2(m_w + 7m_b)]/12$.

**142.**

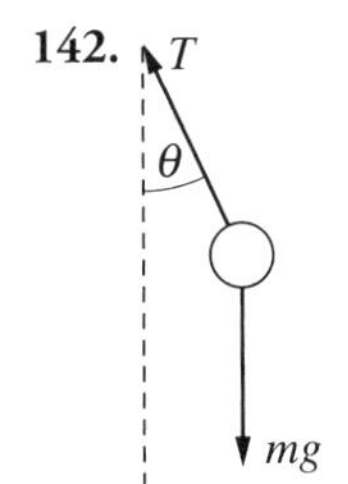


**Figure A7.1**

The external forces on the pendulum ball are shown in Figure A7.1. In the equilibrium situation, $\mathbf{F} = m\boldsymbol{a}$ can be written as $T \sin\theta = m\omega^2 r$ and $T \cos\theta - mg = 0$. Dividing the first expression by the second gives $\tan\theta = \omega^2\tau/g$. Solving for $\theta$, we get 0.55°.

**143.** We note $\Sigma F = ma = m\omega^2 r$ inward along the radial direction. Thus, the scale supplies a force $(mg - mw^2r)$ to support the man. When the earth stops rotating, the force will increase to $mg$.

The ratio we need is $mw^2r/mg = (1.16 \times 10^{-5} \times 2\pi)^2(6.37 \times 10^6)/9.8 = 3.5 \times 10^3$, giving a change of 0.35 percent. At the north pole, the man will be on the axis, so $r = 0$ and the change is zero.

**144.** For just not slipping the friction force, $f = \mu F_N = \mu mg$ supplies the centripetal force, so $\mu mg = m\omega^2 r$, which with $\omega = 45$ rev/min $= 4.71$ rad/s; then $r = 0.25$ m gives $\mu = 0.566$.

**145.** $I = \frac{1}{2}MR^2$

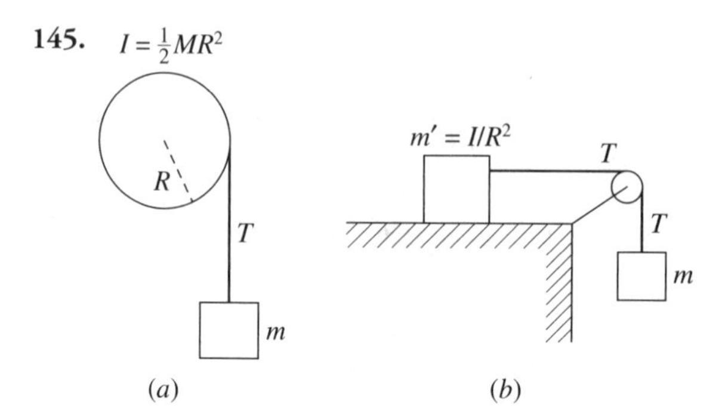


**Figure A7.2**

The situation is depicted in Figure A7.2($a$). We choose downward as positive for the block and for consistency, clockwise as positive for the wheel. We have two dynamical equations. For the block:

$$mg - T = ma \tag{1}$$

where $m = 1.2$ kg and thus $mg = 11.8$ N. Our second equation is $\tau = I\alpha$, with $\tau = TR$, or $TR = I\alpha$. Multiplying both sides by $R$, $TR^2 = IR\alpha$; noting that $a = R\alpha$, we have $TR^2 = Ia$ or

$$T = (I/R^2)a \tag{2}$$

In this form, the equation resembles that of a block on a horizontal frictionless surface, having a mass $(I/R^2)$ and being pulled by a horizontal force $T$. The entire question then resembles the question of two blocks attached by a cord over a frictionless pulley depicted in Figure A7.2($b$). We note that $I = (25 \text{ kg})(0.30 \text{ m})^2 = 2.25 \text{ kg} \cdot \text{m}^2$, $R = 0.40$ m, and $I/R^2 = 14.1$ kg. To obtain the acceleration, $a$, we add Eqs. (1) and (2), eliminating the tension $T$, and get $mg = (m + I/R^2)a$. Substituting numerical values, $11.8 \text{ N} = (1.2 \text{ kg} + 14.1 \text{ kg})a$, or $a = 0.77 \text{ m/s}^2$. Substituting into Eq. (2), we get $T = (14.1 \text{ kg})(0.77 \text{m/s}^2) = 10.9$ N.

**146. (A)** Choose left and counterclockwise as positive. Write $F = ma = 20 + f = 4a$, with $f$ being the friction force at the floor. From $\tau = I\alpha = (20 - f)(0.10) = 0.02(a/0.10)$, and the $F = ma$ equation, we find $a$ to be 6.7 m/s$^2$. Substituting back into either equation, we get $f =$ 6.8 N to the left.

**(B)** In the case $f = 0$ in Question 146(A), the first equation gives $a = 5.0$ m/s$^2$. Note that $a = \alpha r$ is not applicable when slippage occurs.

**147.** Along the incline, $F = ma$ yields $ma = mg \sin 37° - T - 30$, while for the wheel $\tau = I\alpha$ becomes $rT = I\alpha$. Using $a = \alpha r$, we solve $F = ma$ to obtain $T = mg \sin 37° - 30 - m\alpha r$, which is placed into the torque equation, yielding $\alpha = [r(mg \sin 37° - 30)]/(I + mr^2)$. Inserting values for $m$, $r$, $g$, and $I$, we get $\alpha = 1.2 \text{ rad/s}^2$.

**148.** $\text{KE}_1 = \frac{1}{2} I\omega_1^2$ $\quad$ $120 \text{ rev/min} = 4\pi \text{ rad/s} = \omega_1$ $\quad$ $\text{KE}_1 = \frac{1}{2}(900)(4\pi)^2 = 71 \text{ kJ}$

$90 \text{ rev/min} = 3\pi \text{ rad/s} = \omega_2$ $\quad$ $\text{KE}_2 = \frac{1}{2} I\omega_2^2 = \frac{1}{2}(900)(3\pi)^2 = 40 \text{ kJ}$

$\text{KE}_1 - \text{KE}_2 = 31 \text{ kJ}$ $\quad$ loss of energy = work done by brake

**149.** Let $U_g$ = PE of gravity; $K_r$ = rotational KE. The velocity of mass $m$ is $3bw$. To find $w$, equate the $U_g$ lost by the three masses to $K_r$ about pivot point so that $U_g = 2mbg + (3m)(2b)(g) + (m)(3b)(g) = K_r = (Iw^2)/2$. About the pivot point $I$ is $2mb^2 + (3m)(2b)^2 + (m)(3b)^2 = 23mb^2$ so that $w = (22g/23b)^{1/2} = (3.06/b^{1/2})$ rad/s. From this, find $v = 3b\omega = 9.2b^{1/2}$ m/s.

**150.** The rotational and translational KE of the ball at the bottom will be changed to gravitational PE when the sphere stops. We therefore write

$$\left(\frac{1}{2}Mv^2 + \frac{1}{2}I\omega^2\right)_{\text{start}} = (Mgh)_{\text{end}}$$

But for a solid sphere, $I = \frac{2}{5}Mr^2$. Also, $\omega = v/r$. The equation becomes

$$\frac{1}{2}Mv^2 + \frac{1}{2}\left(\frac{2}{5}\right)(Mr^2)\left(\frac{v}{r}\right)^2 = Mgh \qquad \text{or} \qquad \frac{1}{2}v^2 + \frac{1}{5}v^2 = (9.8 \text{ m/s}^2)h$$

Using $v = 20$ m/s gives $h = 28.6$ m. Note that the answer does not depend upon the mass of the ball or the angle of the incline.

**151.** At the top, the disk has translational and rotational KE, plus its GPE relative to the point 18 cm lower. At the final point, the GPE has been transformed to more KE of rotation and translation. We therefore write, with $h = 18$ cm,

$$(\text{KE}_t + \text{KE}_r)_{\text{start}} + Mgh = (\text{KE}_t + \text{KE}_r)_{\text{end}} \qquad \frac{1}{2}Mv_0^2 + \frac{1}{2}I\omega_0^2 + Mgh = \frac{1}{2}Mv_f^2 + \frac{1}{2}I\omega_f^2$$

For a solid disk, $I = \frac{1}{2}Mr^2$. Also, $\omega = v/r$. Substituting these values and simplifying gives

$$\frac{1}{2}v_0^2 + \frac{1}{4}v_0^2 + gh = \frac{1}{2}v_f^2 + \frac{1}{4}v_f^2$$

But $v_0 = 0.80$ m/s and $h = 0.18$ m. Substitution gives $v_f = 1.7$ m/s.

**152.** Net work $= \Delta\text{KE} = (\frac{1}{2}I\omega^2 - 0)$; $W = Fs = (40 \text{ N})(3 \text{ m}) = 120$ J. $\omega = 2\pi f = 4\pi$ rad/s. Thus, $120 \text{ J} = \frac{1}{2} I (4\pi \text{ rad/s})^2$; $I = 1.5 \text{ kg} \times \text{m}^2$.

**153.** Use energy conservation; first find the final speed of mass from $x = \bar{v}t = (v + v_0)(t/2)$ to obtain the final mass velocity $v$ = 0.167 m/s. Then, $U_g$ lost by mass = $K_r + K$, or $mgh = (Iw^2)/2 + (mv^2)/2$; $0.05(9.8)(1.0) = 0.5I(0.167/0.2)^2 + 0.5(0.05)(0.167)^2$. Then, solving for $I$ gives 1.4 kg · m².

**154.** $K_r = (I\omega^2)/2$ is to supply the needed work, with $I = \frac{1}{2}mR^2$. 100 hp for 10 min equals (746 W/hp)(100 hp)(600 s) = $4.48 \times 10^7$ J and $K_r = 0.5(0.5\,m \times 0.50^2)(300 \times 2\pi)^2 = 2.22 \times 10^5 m$; therefore, $m$ = 202 kg.

**155. (A)** The oscillation period for a physical pendulum is given by

$$T = 2\pi\sqrt{\frac{I}{mgD}} = 2\pi\sqrt{\frac{\frac{1}{3}Ml^2}{mg(\frac{l}{2})}} = 2\pi\sqrt{\frac{2l}{3g}}$$

**(B)** The period of a simple pendulum of length $L$ is given by $2\pi\sqrt{L/g}$. Therefore, the length of a simple pendulum with the oscillation period $T_A$ is $L = 2l/3$.

**156.** The ring is shown in Figure A7.3, with the knife edge at point $A$. We must find the period $T_1$, of small oscillations in the plane of the paper. Taking the origin of a coordinate system at $O$, the equilibrium position of the ring's center, with the positive $z$ axis emerging toward the viewer, the moment of inertia $I_{zO} = MR^2$. By the parallel-axis theorem, the moment of inertia about the knife edge is given by $I_{zA} = I_{zO} + MR^2 + 2MR^2$.

$$\text{So}\quad T = 2\pi\sqrt{\frac{2}{mgD}} = 2\pi\sqrt{\frac{2MR^2}{mgR}} = 2\pi\sqrt{\frac{2R}{g}}$$

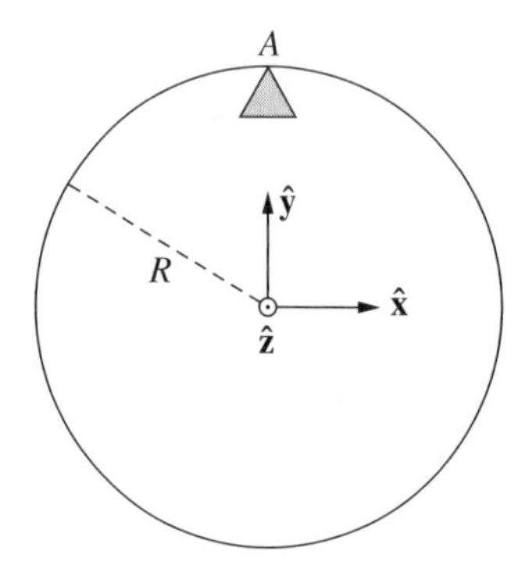


**Figure A7.3**

**157.** $I_1\omega_1 = I_2\omega_2$ $\quad 1.33(1.9) = 0.48\omega_2$ $\quad \omega_2 = 5.3$ rev/s

**158.** Since the line of action of the gravitational force on the satellite always passes through the earth, the angular momentum of the satellite about the earth must be conserved. Thus, $mv_a r_a = mv_p r_p$ or $v_p/v_a = r_a/r_p$.

**159.**

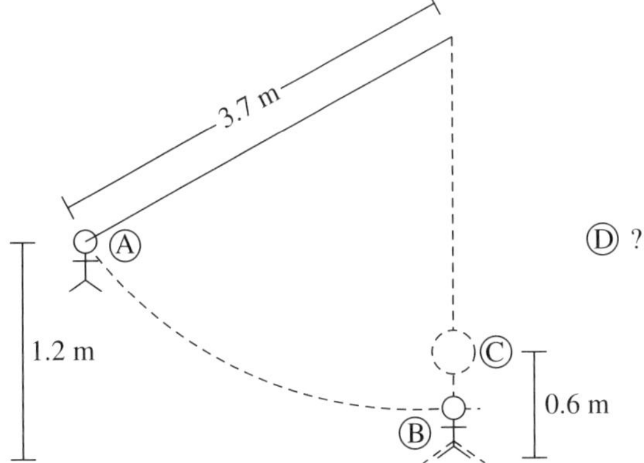


**Figure A7.4**

First find the speed at bottom using energy conservation.

$$\text{PE}_A + \cancel{\text{KE}}_A^0 = \cancel{\text{PE}}_B^0 + \text{KE}_B$$

$$\cancel{m}gh_A = \frac{1}{2}\cancel{m}V_B^2$$

$$V = \sqrt{2gh_A} = \sqrt{2\left(10\frac{\text{N}}{\text{kg}}\right)(1.2\text{ m})} = 4.9\,\frac{\text{m}}{\text{s}}$$

Assume she maintains this speed after she stands up. Do energy conservation again.

$$\text{PE}_C + \text{KE}_C = \text{PE}_D + \cancel{\text{KE}}_D^0$$

$$\cancel{m}gh_C = \frac{1}{2}\cancel{m}V_C^2 = \cancel{m}gh_D$$

$$gh_C + \frac{1}{2}V_C^2 = h_D$$

$$= \frac{\left(10\frac{\text{N}}{\text{kg}}\right)(0.6\,\text{m}) + \frac{1}{2}\left(4.9\,\frac{\text{m}}{\text{s}}\right)^2}{10\,\frac{\text{N}}{\text{kg}}} = 1.8\,\text{m above the lowest position}$$

**160.** No external torque is exerted on the falling sand or disk, so angular momentum is conserved. $I\omega_0 = (I + Mb^2)\omega$, and thus, $\omega = \omega_0/(1 + Mb^2/I)$.

## Chapter 8: Simple Harmonic Motion

**161. (A)** $k = \frac{\Delta F}{\Delta x} = \frac{(0.020\text{ kg})(9.8\text{ m/s}^2)}{0.07\text{ m}} = 2.8\text{ N/m}$

**(B)** $T = 2\pi\sqrt{\frac{m}{k}} = 2\pi\sqrt{\frac{0.050\text{ kg}}{2.8\text{ N/m}}} = 0.84\text{ s}$

**162.** The spring constant is $k = 9/0.05 = 180$ N/m, and so

$$T = 2\pi\sqrt{\frac{m}{k}} = 2\pi\sqrt{\frac{27/9.8}{180}} = 0.78 \text{ s}$$

**163.** We are given frequency $v$ and amplitude $R: f = 2$ Hz; $\omega = 2\pi v = 4\pi$ rad/s; $R = 0.008$ m. Then we have $v_{max} = \omega R = 4\pi(1.008) = 0.101$ m/s and $a_{max} = w^2 R = 16\pi^2(0.08) = 1.264$ m/s$^2$. From Newton's second law, $F_{max} = ma_{max} = (0.5 \text{ kg})(12.6 \text{ m/s}^2) =$ 0.63 N.

**164.** The period varies as the square root of the mass, added mass = (4 – 1)(250 grams) = 750 grams.

**165.** The maximum downward acceleration of the washer will be that for free fall, $g$. If the piston has a greater downward acceleration, the washer will lose contact.

In SHM, the acceleration is given in terms of the displacement and the frequency by $a = -4\pi^2 f^2 x$. With the upward direction chosen as positive, the largest downward (most negative) acceleration occurs for $x = +x_0 = 0.07$ m, that is, $a_0 = 4\pi^2 f^2(0.07 \text{ m})$. The washer will separate from the piston when $a_0$ first becomes equal to $g$. Therefore, the critical frequency is given by

$$4\pi^2 f_c^2 (0.07 \text{ m}) = 9.8 \text{ m/s}^2 \qquad \text{or} \qquad f_c = \frac{1}{2\pi}\sqrt{\frac{9.8}{0.07}} = 1.9 \text{ Hz}$$

**166. (A)** $A = 0.15$ m; $\omega = \sqrt{\frac{k}{m}} = 10$ rad/s; $f = \frac{\omega}{2\pi} = \frac{5}{\pi}$ Hz; $T = \frac{1}{f} = \frac{\pi}{5}$ s

**(B)** $\frac{1}{2}mv^2 + \frac{1}{2}kx^2 = \frac{1}{2}kA^2$ from conservation of mechanical energy, $\frac{1}{2}mv^2 = \frac{1}{2}k(A^2 - x^2) = 2.5$ J.

**167.** Because $2\pi f = (g/L)^{1/2}$, frequency on the moon is $1/(6^{1/2}) = 0.408$ times the frequency on the earth. The number of seconds ticked out will be 0.408(60) = 24.5 s.

**168.** For a pendulum, $2\pi f = [(MgL)/I]^{1/2}$ where $L$ is the distance from the axis of rotation to the center of mass, $I$ is the moment of inertia about the axis of rotation, and $M$ is the mass. For this case, $L = 0.40$ m and $I = I_{cm} + M(0.40)^2$ from the parallel-axis theorem. Since $L_{cm} = [M(1.0)^2]/12$, $I = 0.243M$, and $f = 0.64$ Hz. Similarly, for $L = 0.0010$ m, $I = 0.0833M$, and $f = 0.055$ Hz.

**169.** The spring constant and the mass are independent of location, and so $\omega = 2\pi f = (k/m)^{1/2}$ is also independent of location. Because of the fact that the frequency of a spring system is independent of the initial constant force, the change in initial stretch due to the gravitational change is of no importance; $f = 0.40$ Hz.

## Chapter 9: Hydrostatics

**170. (A)** The pressure to the right of the piston is (30 + 101) kPa, so the compressive force is $(131 \times 10^3 \text{ Pa})(0.50 \times 10^{-4} \text{ m}^2) = 6.55$ N; then dividing by $k = 60$ N/m, we have 10.9 cm.

**(B)** The force is now due to the gauge pressure acting on the piston area; the spring will compress $3.0 \times 10^4(5 \times 10^{-5})/60 = 2.5$ cm.

**171.** First find the pressure at the top of the mercury pool. For a point below the surface of the mercury this may be regarded as a source of external pressure $p_{ext}$. Thus,

$$p_{ext} = \rho_{water} g h_{water} = (1.0 \times 10^3 \text{ kg/m}^3)(9.8 \text{ m/s}^2)(1.2 \text{ m}) = 12 \text{ kPa}$$

The pressure $p_{int}$ exerted by the mercury column itself is found in the same manner:

$$p_{int} = \rho_{merc} g h_{merc} = (13.6 \times 10^3 \text{ kg/m}^3)(9.8 \text{ m/s}^2)(0.30 \text{ m}) = 40 \text{ kPa}$$

The total pressure at the bottom is thus 52 kPa.

**172.** At points $A$ and $B$, pressures must be equal, since the fluid is not moving. Therefore, $P = P_a + \rho g h = 100 \text{ kPa} + [(13.6)(9.8)(0.12)] \text{ kPa} = 116 \text{ kPa}$.

**173.** The *total* pressure in the tank will be the pressure of the atmosphere (about $1.0 \times 10^5$ Pa) plus the pressure due to the piston and weights.

$$p = 1.0 \times 10^5 \text{ N/m}^2 + \frac{(20)(9.8)\text{ N}}{8 \times 10^{-4} \text{ m}^2} = 1.0 \times 10^5 \text{ N/m}^2 + 2.45 \times 10^5 \text{ N/m}^2$$

$$= 3.45 \times 10^5 \text{ N/m}^2 = 345 \text{ kPa}$$

A pressure gauge on the tank would read the difference between the pressure inside and outside the tank: gauge reading $= 2.45 \times 10^5 \text{ N/m}^2 = 245$ kPa. It reads the pressure due to the piston and weights.

**174. (A)** Pascal's principle says that the pressure change is uniformly transmitted throughout the oil, so $\Delta p = F_1/A_1 = F_2/A_2$, where $F_1$ and $F_2$ are the forces on the small and on the large pistons, respectively, and $A_1$ and $A_2$ are the respective areas. Thus,

$$\frac{F_1}{\pi(2^2/4)} = \frac{1500 \times 9.8}{\pi(30^2/4)}$$

Multiplying both sides by $\pi/4$ and solving for $F_1$, we obtain

$$F_1 = \frac{(1500)(9.8)(2^2)}{30^2} = 65 \text{ N}$$

**(B)** $$\Delta p = \frac{F_1}{A_1} = \frac{65}{\pi(2^2/4)} = 21 \text{ N/cm}^2 = 210 \text{ kPa}$$

**175.** The pressures at points $H_1$ and $H_2$ are equal, since they are at the same level in a single connected fluid. Therefore,

$$\text{pressure at } H_1 = \text{pressure at } H_2$$

$$\begin{pmatrix}\text{pressure due to}\\ \text{left piston}\end{pmatrix} = \begin{pmatrix}\text{pressure due to } F\\ \text{and right piston}\end{pmatrix} + (\text{pressure due to 8 m of oil})$$

$$\frac{(600)(9.8)\text{ N}}{0.08 \text{ m}^2} = \frac{F}{25 \times 10^{-4} \text{ m}^2} + (8 \text{ m})(780 \text{ kg/m}^3)(9.8 \text{ m/s}^2)$$

Solving for $F$ gives 31 N.

**176.** The block is in equilibrium under the action of three forces—the weight $w = 71.2$ N, the tension $T$, and the buoyant force $B$, with $B = w + T$. We can determine $B$, since $B = \rho_L g V_B$, where $\rho_L$ is the density of water and $V_B$ is the volume of the totally immersed block; $w = \rho_B g V_B$, where $\rho_B$ is the density of the block. Then $w/B = \rho_B/\rho_L = (\text{sp gr})_B = 0.75$, so $B = w/0.75 = 94.9$ N. Finally, from our equilibrium equation, 94.9 N = 71.2 N + $T$, or $T = 24$ N.

**177.** The desired density is given by $\rho = m/V$. But since the volume $V$ of the ball is also the volume of displaced water, the buoyant force is given by $B = \rho_{\text{water}} g V$. Thus,

$$\rho = \frac{(mg)\rho_{\text{water}}}{B}$$
$$= \frac{(0.096 \text{ N})(1\times 10^3 \text{ kg/m}^3)}{(0.096 - 0.071) \text{ N}}$$
$$= 3840 \text{ kg/m}^3$$

**178.** $mg = \rho_{\text{ice}} v_{\text{ice}} g$

$$F_B = \rho_{\text{water}} V_{\text{submerged}} g$$
$$mg = F_B$$

So $\rho_{\text{ice}} v_{\text{ice}} g = \rho_{\text{water}} v_{\text{sub}} g$

$$\frac{v_{\text{sub}}}{v_{\text{ice}}} = \frac{\rho_{\text{ice}}}{\rho_{\text{water}}}$$

fraction submerged $= \rho_{\text{ice}}/\rho_{\text{water}} = \frac{917}{1025} = 0.89$

**179.** The woman's weight plus the block's weight must be equal to the buoyant force on the just barely submerged block, $50g + 850Vg = 1000Vg$, which leads to $V = 0.33$ m$^3$.

**180.** The tension in the thread is equal to the weight of the metal less the buoyant force. The buoyant force will be $(1 \times 10^{-6} \text{ m}^3)(1000 \text{ kg/m}^3)g = 10^{-3}g$ N, where $g = 9.8$ m/s$^2$. The weight of the metal is $3 \times 10^{-3}g$ N. Therefore, the thread exerts an upward force of $2 \times 10^{-3}g$ N. Hence the scale supports the total weight less the tension in the thread. Therefore, the apparent weight read by a scale will be $(23 - 2)\,g$, or 0.206 N.

Equivalently, we can get the result by noting that if the water exerts an upward buoyant force of $10^{-3}g = 9.8 \times 10^{-3}$ N on the metal, by Newton's third law the metal exerts a like force downward on the water. Thus, the scale balances the weight, 0.196 N, plus the downward force of the metal, 0.0098 N, for a total of 0.206 N.

**181. (A)** The block is in equilibrium, so $\rho_B V_B g = \text{BF} = \rho_0 (2V_B/3)g$. Since $V_B = 4.22 \times 10^{-7}$ m$^3$ and $\rho_0 = 800$ kg/m$^3$, BF $= 2.21\times 10^{-3}$ N.

**(B)** Since BF $= \rho_B V_B g$, we find $\rho_B = 2\rho_0/3 = 533$ kg/m$^3$.

**182. (A)** The buoyant force equals the weight of the displaced liquid, BF $= \rho_{\text{oil}} V_{\text{Cu}} g$, with the volume $V = 3.38\times 10^{-6}$ m$^3$ and $\rho = 820$ kg/m$^3$. Thus, BF = 0.027 N.

**(B)** The force acting on the block are the tension $T$ up, BF up, and the block weight $\rho_{Cu}V_{Cu}g = 8920(3.38\times10^{-6})(9.8) = 0.295$ N acting downward. Then $T = 0.295 - 0.027 = 0.268$ N.

**183.** By equilibrium and Archimedes' principle, $500g = 1.29gV$, or $V = 388$ m$^3$.

## Chapter 10: Hydrodynamics

**184.** If the radius decreases from $r$ to $r/4$, the cross-sectional area decreases from $\pi r^2$ to $\pi r^2/16$. From the continuity equation, assuming an incompressible fluid, the velocity would have to *increase sixteen-fold* at the constriction.

**185.** Use Bernoulli's equation: $p_1 + \rho g h_1 + (\rho v_1^2)/2 = p_2 + \rho g h_2 + (\rho u_2^2)/2$. Using known values and $h_2 - h_1 = 1$ m, one obtains $p_2 = 6.7$ kPa. For no flow, $v_1 = v_2 = 0$, and so $p_2 = p_1 + \rho g(h_1 - h_2) = 8.2$ kPa.

**186.** From the equation of continuity we have, since $\rho =$ constant,
**(A)** $\pi(0.25)^2 v = 3.0$, so $v = 15.3$ cm/s and
**(B)** $\pi(0.40)^2 v = 3.0$, from which $v = 6.0$ cm/s

**187.** Referring to Figure 187.1, we conclude that $p_5 = p_{atm}$. Neglecting fluid friction and assuming that the flow is steady, we can employ Bernoulli's equation to evaluate the fluid pressure at various other locations:

$$p_M - p_N = \frac{1}{2}\rho(v_N^2 - v_M^2) + \rho g(y_N - y_M)$$

Since the cross-sectional area of the container is the same at points 2, 4, and 5, the flow speeds must also be the same: $v_2 = v_4 = v_5$. The elevations are also the same: $y_2 = y_4 = y_5$. Bernoulli's equation therefore implies that $p_2 = p_4 = p_5$. Comparing point 1 with a point at the top surface of the liquid, and assuming that the flow speed is negligible at both locations, we find that $p_1 = p_{atm} + pgh$. Application of Bernoulli's equation to points 1 and 5 shows that $v_5 = \sqrt{2gh}$. Since the cross-sectional area $A_3$ is less than $A_5$, we must have $v_3 > v_5$. If we assume that $A_3 = \frac{1}{2}A_5$, we find that $v_3 = 2v_5$. Bernoulli's equation then implies that $p_3 = p_5 - 3\rho gh$. In summary, we have $p_1 = p_{atm} + \rho gh$, $p_2 = p_4 = p_5 = p_{atm}$, and (assuming $A_3 = \frac{1}{2}A_5$) $p_3 = p_{atm} - 3\rho gh$

**188.** Apply Bernoulli's equation to the differences $\Delta v^2$, $\Delta p$, and $\Delta y$ between the locations $t$ and $b$ in Figure 188.1. Since the tank is large and the pipe is small, it may be assumed that the water has negligible speed until it is actually in the outlet pipe. Bernoulli's equation, $\Delta p + \frac{1}{2}\rho\ \Delta v^2 + \rho g\Delta y = 0$, can be solved by noting that $\Delta v^2 = v_b^2 - 0$ and $\Delta p = 0$ (both the jet and the top surface are at atmospheric pressure). Bernoulli's equation thus becomes $\frac{1}{2}\rho\ \Delta v^2 + \rho g\ \Delta y = 0$, or $v_b^2 = -2g\ \Delta y$, and so $v_b = \sqrt{2g(-\Delta y)}$.
The outlet speed is the free-fall speed—this is *Torricelli's theorem*.
Inserting the numerical values gives $v_b = \sqrt{2\times 9.8\text{ m/s}^2 \times 8.0\text{ m}} = 13$ m/s.

**189.** From Torricelli's theorem, $v = \sqrt{2gd}$, where $v$ is the horizontal velocity of efflux from the barrel and $d$ is the depth of interest. The efflux is horizontal, so the time to hit the ground is given by $h = \frac{1}{2}gt^2$, or $t = \sqrt{2h/g}$ and $R = vt = \sqrt{2gd}\sqrt{2h/g} = 2\sqrt{dh}$. Then $R^2 = 4dh$ and $d = R^2/(4h)$.

**190.** Using Torricelli's theorem for the escape speed, we have for the volume flow

$$vA = \sqrt{2gh}\,A = \sqrt{2(9.8\times10^3 \text{ mm/s}^2)(20\times10^3 \text{ mm})}\ (1 \text{ mm}^2)$$
$$= 19800 \text{ mm}^3/\text{s} \quad 19.8 \text{ mL/s}$$

**191.** By Torricelli's theorem, the water leaves the hole at speed $v = \sqrt{2gy}$. Each element of the water stream follows the trajectory of a particle launched horizontally at the same speed and elevation. The "fall time" $t$ satisfies the equation $\frac{1}{2}gt^2 = h - y$, so that $t = \sqrt{2(h-y)/g}$. The horizontal distance traveled during the fall is therefore given by

$$R = vt = \sqrt{2gy}\sqrt{\frac{2(h-y)}{g}} = 2\sqrt{y(h-y)} \qquad \textbf{(1)}$$

**192.** To find $v$ equate $KE_{bottom}$ to $PE_{top}$, giving $v = (2gh)^{1/2} = 7.0 \text{ m/s} = 700 \text{ cm/s}$. The flow rate is $Av$, so in 1 min we have $(0.75 \text{ cm}^2)(700 \text{ cm/s})(60 \text{ s}) = 3.15\times10^4 \text{ cm}^3 = 31.5 \text{ L}$.

**193.** We denote the initial speed by $v_0$ and the initial cross-sectional area by $A_0$. After the freely falling stream has descended a distance $h$, its speed $v_1 = \sqrt{v_0^2 + 2gh}$. Under steady-flow conditions, the mass fluxes at the locations are equal: $\rho_0 v_0 A_0 = \rho_1 v_1 A_1$. Since the water is effectively incompressible, $\rho_0 = \rho_1$ and therefore $A_1 = (v_0/v_1)A_0$. Inserting numerical values, we find $v_1 = \sqrt{(3.0)^2 + 2(9.8)(0.50)} = 4.34$ m/s. Then $A_1 = (3.4/4.34)(1.0 \text{ cm}^2) = 0.69 \text{ cm}^2$.

**194.** Applying Bernoulli's equation between the top of the tank and ground level,

$$p_{atm} + \tfrac{1}{2}\rho(0^2) + \rho g(h_0 + H) = p_{atm} + \tfrac{1}{2}\rho v_3^2 + \rho g(0)$$

whence $v_3 = \sqrt{2g(h_0 + H)} = \sqrt{2(9.8)(9)} = 13.2$ m/s (as though in free fall all the way). This result holds even if we lose laminar flow in the stream to the ground, since it follows from conservation of energy for each droplet of water.

## Chapter 11: Temperature and Thermal Expansion

**195.** $\alpha = \dfrac{1}{L_0}\dfrac{\Delta L}{\Delta T} = \dfrac{0.091\times10^{-2}\text{ m}}{(3 \text{ m})(60 \text{ K})} = 5.1\times10^{-6}\text{K}^{-1}$

**196.** $\Delta L = \alpha L_0 \Delta T$. For our case, $L_0 = 29.930$ in and $\Delta L = 0.070$ in. Then $\Delta T = 0.070 \text{ in}/[(1.2\times10^{-5}\,°\text{C}^{-1})(29.930 \text{ in})] = 195°\text{C}$. Finally, $t_C = 15°\text{C} + 195°\text{C} = 210°\text{C}$.

**197.** We want $\Delta L/L = \pm10^{-6} = \alpha\Delta T = 12\times10^{-6}\Delta T$, so $\Delta T = \pm0.083°\text{C}$.

**198.** We let $I$ stand for the iron ball and $B$ stand for the brass plate. $L_I = 6$ cm and $L_I - L_B = 0.001$ cm at $t = 30°$C. Since the brass plate expands uniformly, the hole must expand in the same proportion. Then heating both the ball and the plate leads to increases in the diameters of the ball and the hole, with the hole increasing more, since $\alpha_B > \alpha_I$. We require

$\Delta L_B - \Delta L_I = 0.001$ cm. $\Delta L_B = \alpha_B L_B \Delta t$; $\Delta L_t = \alpha_I L_I$. We can approximate $L_B$ in this formula by 6 cm = $L_I$. Then

$$\Delta L_B - \Delta L_I = (\alpha_B - \alpha_I) L_I \Delta t = 0.001 \text{ cm}$$

or

$$[(1.9\times10^{-5}\,°\text{C}^{-1}) - (1.2\times10^{-5}\,°\text{C}^{-1})](6 \text{ cm})\,\Delta t = 0.001 \text{ cm}$$

Solving, $\Delta t = 23.8°\text{C}$, and finally $t = 30°\text{C} + 23.8°\text{C} = 53.8°\text{C}$.

**199.**

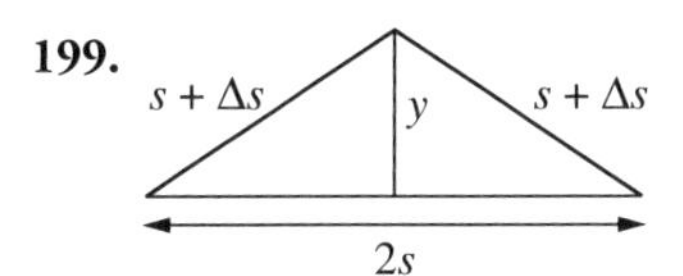


**Figure A11.1**

As indicated in Figure A11.1, we let the initial length be $2s$ and the final total length be $2(s + \Delta s)$. The height of the center of the buckled rail is denoted by $y$. We have $\Delta s = \alpha s\ \Delta T$. By the Pythagorean theorem,

$$\begin{aligned} y &= \sqrt{(s+\Delta s)^2 - s^2} \\ &= \sqrt{2s\Delta s + (\Delta s)^2} \\ &= s\sqrt{2\alpha\,\Delta T + (\alpha \Delta T)^2} \end{aligned}$$

With $s = 15.0$ m, $\alpha = 12\times10^{-6}\ \text{K}^{-1}$, and $\Delta T = 50$ K, we obtain

$$\begin{aligned} y &= (15.0 \text{ m})\sqrt{(12\times10^{-4}) + (6\times10^{-4})^2} \\ &= 0.52 \text{ m} \end{aligned}$$

**200.** **(A)** The 88.42-cm mark on the aluminum rod is really at a greater distance from the zero position than indicated because of the increase in temperature, $\Delta t = 30°\text{C}$. This increased length is $\Delta L = \alpha_{\text{Al}} L_{\text{Al}} \Delta t = (2.55\times10^{-5}\,°\text{C}^{-1})(88.42 \text{ cm})(30°\text{C}) = 0.068$ cm.

**(B)** At 35°C, the measuring rod at the 88.42-cm mark is, from part **(A)**, actually 88.49 cm long. It thus measures an 88.49-cm length of steel at 35°C.

**201.** $\Delta L = \alpha L\ \Delta t = (6.7\times10^{-6})(150)(15° - 70°) = (6.7\times10^{-6})(150)(-55°)$

$\Delta L = -0.055$ ft

The tape is 0.055 ft too short, and the reading obtained is too large.

**202.** For a temperature change from 20°C to –15°C, we have $\Delta T = -35°\text{C}$. Then,

$$\frac{\Delta L}{L_0} = \alpha \Delta T = (1.2\times10^{-5}\,°\text{C}^{-1})(-35°\text{C}) = -4.2\times10^{-4} = -0.042\%$$

**203.** At 20°C, $T_0 = 2\pi(L_0/g)^{1/2} = 1$ s. The brass elongates to $L = L_0[1 + (19.3\times10^{-6})\times(10°\text{C})]$, so the new period $T = 2\pi(L/g)^{1/2} = T_0(1 + 1.93\times10^{-4})^{1/2} = 1.0001T_0$. At 30°C, the clock loses $(T - T_0)$(1 week) = $(10^{-4}\text{ s})(60 \times 60 \times 24 \times 7) = 60$ s. (Note that an increase in period means that the clock ticks off fewer seconds and hence is slow.)

**204.** The volume of glycerin that overflows is equal to the difference in the volume expansion of glycerin $\Delta V_g$ and the volume expansion of aluminum $\Delta V_a$.

$$\Delta V_g = \beta_g V \Delta t \qquad \Delta V_a = 3\alpha V \Delta t$$

$$\Delta V_g - \Delta V_a = \beta_g V \Delta t - 3\alpha V \Delta t = (\beta_g - 3\alpha) V \Delta t$$

$\alpha$ equals $2.55 \times 10^{-5}\,°\text{C}^{-1}$, $\beta_g$ equals $5.3 \times 10^{-4}\,°\text{C}^{-1}$.

$$\Delta V_g - \Delta V_a = [(5.3 \times 10^{-4}) - 3(2.55 \times 10^{-5})](300)(100° - 20°)$$

$$= (4.535 \times 10^{-4})(27\,000) = 12.2 \text{ mL}$$

Thus, 12.2 mL of glycerin overflows.

**205.** Let $\rho_0$ = density of mercury at 0°C, $\rho_1$ = density of mercury at 50°C, $V_0$ = volume of $m$ kg of mercury at 0°C, $V_1$ = volume of $m$ kg of mercury at 50°C. By conservation of mass, $m = \rho_0 V_0 = \rho_1 V_1$, from which

$$\rho_1 = \rho_0 \frac{V_0}{V_1} = \rho_0 \frac{V_0}{V_0 + \Delta V} = \rho_0 \frac{1}{1 + (\Delta V / V_0)}$$

But

$$\frac{\Delta V}{V_0} = \beta \Delta T = (1.82 \times 10^{-4}\,°\text{C}^{-1})(50°\text{C}) = 0.0091$$

Substitution gives

$$\rho_1 = (13{,}600 \text{ kg/m}^3) \frac{1}{1 + 0.0091} = 13{,}500 \text{ kg/m}^3$$

## Chapter 12: Heat and Calorimetry

**206.** Consider mass $m$ of water falling.

$$mgy = mc\ \Delta t \qquad gy = c\ \Delta t$$

We express both sides in joules by noting

$$c = 1 \text{ kcal/kg} \cdot \text{K} = 4184 \text{ J/kg} \cdot \text{K}$$

Then

$$9.8(122) = 4184 \Delta t \quad \text{and} \quad \Delta t = 0.29 \text{ K}$$

**207.** $K = \frac{1}{2} m v^2 = mc\ \Delta T$ for the lead bullet. The left side is calculated in joules and converted to calories: $K = 0.0022(150)^2/2 = 24.8 \text{ J} = (24.8/4.184) \text{ cal} = 5.91$ cal. Then $5.91 \text{ cal} = (2.2 \text{ g})(0.031 \text{ cal/g} \cdot °\text{C}) \Delta T$ and $\Delta T = 87°\text{C}$.

**208.** Here some of the initial kinetic energy is carried away by the bullet-block combination. From momentum conservation, the velocity of the system after the collision is given by (2.2 g)(150 m/s) = (50 g + 2.2 g)$V$ and $V$ = 6.32 m/s. $K_f = \frac{1}{2}(0.0522 \text{ kg})(6.32 \text{ m/s})^2 = 1.04 \text{ J} = 0.25$ cal. $\Delta Q$ is now $5.91 - 0.25 = 5.66 \text{ cal} = 2.2(0.031)\ \Delta T'$ and $\Delta T' = 83°\text{C}$.

**209.** Express kinetic energy and heat energy in joules and the mass in kilograms.

$$0.40(\Delta\text{KE}) = Q \qquad 0.40\left(\frac{1}{2}mv_1^2 - \frac{1}{2}mv_2^2\right) = mc\ \Delta t \qquad 0.40\left(\frac{1}{2}v_1^2 - \frac{1}{2}v_2^2\right) = c\ \Delta t$$

$$0.40\left(\frac{1}{2}\times 500^2 - \frac{1}{2}\times 300^2\right) = (0.031)(4186)\Delta t \qquad \text{and} \qquad \Delta t = 247°\text{C}$$

The temperature rise of the bullet is 247°C.

**210.** Potential energy is converted into heat. $U_g = (100m)(9.8)(1.5) = 1470m$ J, where $m$ is in kilograms. This equals (130 J/kg · °C)($m$ kg) $\Delta T$. Solving for $\Delta T$ one obtains 11.3°C. Note that the mass $m$ drops out of the calculation.

**211.** The boy's kinetic energy is changed to heat energy. Set $Q = (mv^2)/2 = [60(25)]/2 =$ 750 J = 179 cal. From $Q = c\rho V\Delta T$, 179 cal = (1.0 cal/g · °C)(0.950 g/cm$^3$)(2.0 cm$^3$)$\Delta T$, where $\Delta T = 94$°C.

**212.** Each second, $\frac{300}{60} = 5.0$ g is heated. Thus $\Delta Q = cm\ \Delta T$ = (1.0 cal/g · °C)(5.0 g/s) (71°C)(4.184 J/cal) = 1.5 kW.

**213.** The heat added is (1.8 kJ/s)$t$ and the heat absorbed to $cm\ \Delta T$ = (4.184 kJ/kg · K) (200 kg)(60 K) = $5.0\times 10^4$ kJ. Equating heats, $t = 2.78\times 10^4$ s = 7.75 h.

**214.** **(A)** The bullet must first be heated to 300°C and then melted. Heat needed = (6 g)(0.20 cal/g · °C) (300 − 0)°C + (6 g)(15 cal/g) = 450 cal, or 1800 J.

**(B)** 1800 J is to be supplied as $K = (mv^2)/2$. 1800 J $= \frac{1}{2}(0.006\text{ kg})v_{\text{min}}^2$ so that $v_{\text{min}} = 790$ m/s.

**215.** The temperature rise $\Delta t$ is obtained from $Q = mc\,\Delta t$, or 2000 kcal = (60 kg) × (0.83 kcal/kg × °C) $\Delta t$. Solving, we have $\Delta t = 40$°C.

**216.**
$$\left(\begin{array}{c}\text{heat lost by}\\ \text{metal block}\end{array}\right) = \left(\begin{array}{c}\text{heat gained}\\ \text{by water}\end{array}\right) + \left(\begin{array}{c}\text{heat gained by}\\ \text{copper calorimeter}\end{array}\right) (mc\,\Delta T)_{\text{metal}}$$
$$= (mc\ \Delta T)_{\text{water}} + (mc\,\Delta T)_{\text{copper}}$$

Note that each term in parentheses is positive as defined. Thus the $\Delta T$ values are all chosen positive. We have

$$(1\text{ kg})c(100°\text{C} - 40°\text{C}) = (0.45\text{ kg})(1.00\text{ kcal/kg}\cdot°\text{C})(40°\text{C} - 20°\text{C})$$
$$+ (0.30\text{ kg})(0.093\text{ kcal/kg}\cdot°\text{C})(40°\text{C} - 20°\text{C})$$

Solving, we get $c = 0.159$ kcal/kg · °C.

**217.** Heat gained = heat lost. (We assume no heat transfer to or from the container.)

$$(50\text{ g})(1.00\text{ cal/g}\cdot{}^\circ\text{C})(t-0^\circ\text{C}) = (250\text{ g})(1.00\text{ cal/g}\cdot{}^\circ\text{C})(90^\circ\text{C}-t)$$

where $t$ is the final equilibrium temperature

$$50t = 22500 - 250t \quad \text{or} \quad 300t = 22500 \quad t = 75^\circ\text{C}$$

**218.** The nail acquired the temperature $t$ of the flame. When the nail was plunged into the cold water, it lost an amount of heat equal to the heat gained by the water. The specific heat of iron can be looked up to be 0.11 cal/g·°C.

$$(10\text{ g})(0.11\text{ cal/g}\cdot{}^\circ\text{C})(t-20^\circ\text{C}) = (100\text{ g})(1.00\text{ cal/g}\cdot{}^\circ\text{C})(20^\circ\text{C}-10^\circ\text{C})$$

Therefore, $t$ – 20°C = (1000/1.1) = 910°C so that the result for the flame temperature is 930°C.

**219.** Heat lost by Ni = heat gained by water + calorimeter.
The specific heat of nickel is 0.106 cal/g·°C,

$$(0.250)(0.106)(120^\circ - t) = [(0.200)(1.00) + 0.020)](t - 10^\circ\text{C})$$

$$3.18 - 0.027t = 0.220t - 2.20 \qquad 0.247t = 5.38 \qquad t = 22^\circ\text{C}$$

**220.** Heat lost = heat gained $c(500)(80-20) = cm(20)$, which yields $m = 1500$ g.

**221.** Heat lost = heat gained is written as $0.11(500)(400 - t) = 0.40(800)(t - 20)$, from which $t = 75.7^\circ$C.

**222.** Here we use $\Delta H$ as the algebraic expression for heat gained by a system. Thus it is positive for heat entering a system and negative when it leaves a system. The heat gained by the can and the ice is equal to the heat lost by the hot water. Assuming that the hot water is sufficient in quantity to melt all the ice, the heat gained by the can and the ice is given by

$$\Delta H_{\text{can \& ice}} = m_{\text{ice}}L_{\text{ice}} + (m_{\text{ice}}c_{H_2O} + m_{\text{can}}c_{\text{Al}})(t_f - t_c)$$

Here $m_{\text{ice}}$ is the mass of ice, $m_{\text{can}}$ is the mass of the can, $t_c$ is the initial temperature of the can and ice, $t_f$ is the final temperature, $L_{\text{ice}}$ is the latent heat of the melting of ice, $c_{H_2O}$ is the specific heat capacity of water, and $c_{\text{Al}}$ is the specific heat capacity of aluminum. The heat lost by the hot water is

$$-\Delta H_{\text{hot water}} = m_{\text{hot water}}\, c_{H_2O}(t_h - t_f)$$

where $m_{\text{hot water}}$ is the mass of the hot water and $t_h$ is its initial temperature. Setting $\Delta H_{\text{can \& ice}} = -\Delta H_{\text{hot water}}$, we have

$$m_{\text{ice}}L_{\text{ice}} + (m_{\text{ice}}c_{H_2O} + m_{\text{can}}c_{\text{Al}})t_f - (m_{\text{ice}}c_{H_2O} + m_{\text{can}}c_{\text{Al}})t_c = m_{\text{hot water}}\, c_{H_2O}t_h - m_{\text{hot water}}\, c_{H_2O}t_f$$

Solving for $t_f$, we find

$$t_f = \frac{m_{\text{hot water}}c_{H_2O}t_h + (m_{\text{ice}}c_{H_2O} + m_{\text{can}}c_{\text{Al}})t_c - m_{\text{ice}}L_{\text{ice}}}{m_{\text{hot water}}\, c_{H_2O} + m_{\text{ice}}c_{H_2O} + m_{\text{can}}c_{\text{Al}}}$$

Inserting the numerical values $c_{H_2O} = 1.00$ cal/g·°C, $c_{Al} = 0.22$ cal/g·°C, and $L_{ice}$ = 79.8 cal/g, we obtain

$$t_f = \frac{600+(48.0+0.44)(0)-(3830)}{(75.0+48.0+0.44)} = 17.6°\text{C}$$

(*Note:* If a negative value had been obtained for $t_f$, it would have shown the incorrectness of our a priori assumption that the hot water was sufficient to melt all of the ice.)

**223.** We assume that the specific heat of coffee is that of water, and use Table 17-1 to get $L_i$. Heat gained = heat lost: $m(80) + 1.0(m)(60) = 1.0(200)(30)$, from which $m = 43$ g.

**224.** The heat of vaporization of water is 540 kcal/kg. Let $T$ be the final temperature. Use $Q = mc\,\Delta t$.

$$\begin{aligned} \text{heat lost} &= \text{heat gained} \\ 0.0061(540)+0.0061(1)(100°-T) &= 0.130(1)(T-20°)+0.125(0.10)(T-20°) \\ 3.294+0.61-0.0061T &= 0.130T-2.60+0.0125T-0.250 \\ 6.754 &= 0.1486T \\ T &= 45.5°\text{C} \end{aligned}$$

## Chapter 13: Heat Transfer

**225.** $Q = \dfrac{k\,(\text{area})(t_h - t_c)(\text{time})}{\text{thickness}} = \dfrac{0.0025(25\times 25)(40)(3600)}{4} = 56$ kcal

**226.** For the same $\Delta T$ and $A$, the $\Delta Q/\Delta t$ is the same in the two materials, so $k_w/L_w = k_b/L_b$, and the wood thickness is $L_w = (0.1/0.8)(8 \text{ cm}) = 1$ cm.

**227.** The heat flow into the box is $\Delta Q/\Delta t = kA(\Delta T/\Delta x) = (0.040)(1.5)(30/0.05) = 36$ W; in an hour the total energy is 130 kJ. This equals $(m)$(80 kcal/kg)(4.184 kJ/kcal), so $m$ = 0.39 kg of ice will melt.

**228.** The rate of heat flow through the glass is (0.80 W/m · K)(1 m$^2$)(20 K/0.005 m) = 3200 W. This is an overestimate, since semistagnant air layers on each side also exist, making the actual temperature difference across the glass much less than 20 K.

**229.**

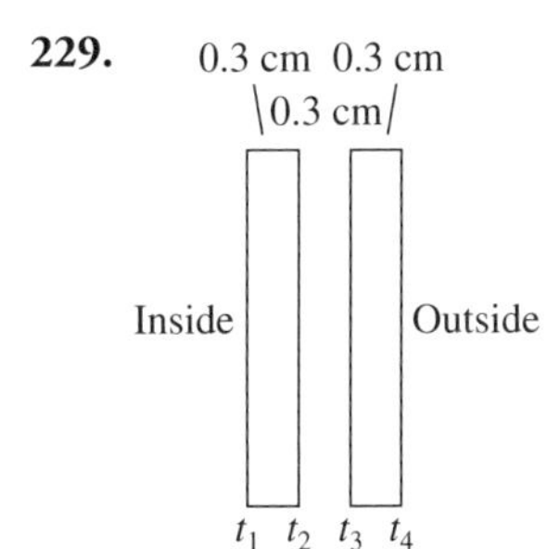


Figure A13.1

A cross section of the window is shown in Figure A13.1; the four surface temperature are labeled, and $t_1 = 20°C$, $t_4 = 0°C$.

$$H_1 = \frac{(2\times10^{-3})(6400)(20-t_2)}{0.30} = 42.7(20-t_2)$$

$$H_2 = \frac{(2\times10^{-4})(6400)(t_2-t_3)}{0.30} = 4.27(t_2-t_3)$$

$$H_3 = \frac{(2\times10^{-3})(6400)(t_3-0)}{0.30} = 42.7t_3$$

Since the three rates of flow must be equal in a steady-state situation, $H_1 = H_2 = H_3$, we have

$$42.7(20-t_2) = 4.27(t_2-t_3) = 42.7t_3 \qquad 10(20-t_2) = (t_2-t_3) = 10t_3$$

yielding

$$t_2 = 11t_3, \quad 20-t_2 = t_3 \Rightarrow 20 = 12t_3, \quad t_3 = 1.67°\text{C} \quad \text{and} \quad t_2 = 18.4°\text{C}$$

Finally,

$$H = H_3 = (42.7)(1.67) = 71 \text{ cal/s}$$

**230.** If the motion of a fluid is caused by a difference in density that accompanies a change in temperature, the current produced is referred to as *natural convection*. When a fluid is caused to move by the action of a pump or fan, the current produced is referred to as *forced convection*. Both types of convection are employed in a common home heating system: Hot water is circulated through "radiators" by forced convection; the warmed air rises by natural convection.

**231.** The surface area of a sphere is $4\pi r^2$. In this case, then, the area is $4\pi(25\times10^{-4}) = 0.01\pi \text{ m}^2$. The power radiated is given by Stefan's law:

$$P = \sigma T^4 A = (5.67\times10^{-8} \text{ W/m}^2\cdot\text{K}^4)(600 \text{ K})^4(0.01\pi \text{ m}^2) = 231 \text{ W}$$

## Chapter 14: Gas Laws and Kinetic Theory

**232.** **(A)** $n = \dfrac{pV}{RT} = \dfrac{(1.52\times10^6)(10^{-2})}{(8.31)(298)} = 6.1 \text{ mol}$

**(B)** The atomic mass of hydrogen is 1.008, so that 1 mol of hydrogen ($H_2$) contains 2.016 g, or $2.016\times10^{-3}$ kg. The density of the hydrogen is then

$$\rho = \frac{nM}{V} = \frac{(6.13)(2.016\times10^{-3})}{10^{-2}} = 1.2 \text{ kg/m}^3$$

**(C)** The atomic mass of oxygen is 16, so that 1 mol of $O_2$ contains 32 g, or $32\times10^{-3}$ kg. The density of the oxygen is then

$$\rho = \frac{nM}{V} = \frac{(6.13)(32\times10^{-3})}{10^{-2}} = 20 \text{ kg/m}^3$$

**233.** For a confined gas,

$$\frac{p_1V_1}{T_1}=\frac{p_2V_2}{T_2} \qquad \frac{1(500)}{300}=\frac{0.5V_2}{270} \qquad V_2=\frac{500(270)}{300(0.5)}=900\ \text{m}^3$$

**234.** $p_0 = p\dfrac{V}{V_0} = (100\ \text{kPa})(4) = 400\ \text{kPa}$

**235.** Since the temperature is fixed, we use Boyle's law to solve the question. The total pressure in the tank has been reduced from 31 to 26 atm. At the latter pressure, the gas originally in the tank would occupy $\frac{31}{26}\ \text{m}^3$. Since 1.00 $\text{m}^3$ remains in the tank, the amount of gas used was $\frac{5}{26}\ \text{m}^3$ at 26 atm. At the same temperature and at atmospheric pressure, this would occupy a volume 26 times as large: 5.00 $\text{m}^3$.

**236.** $p_1 = p_0\dfrac{T_1}{T_0} = [(14.7+24)\text{psi}]\dfrac{333\ \text{K}}{293\ \text{K}} = 44.0\ \text{psi}$ or 29.3 psi gauge pressure

**237.** Use $(P_1V_1)/T_1 = (P_0V_0)/T_0$ to find $T_1 = (P_1/P_0)(V_1/V_0)T_0 = (50)\left(\frac{1}{16}\right)(273) = 853\ \text{K}$.

**238.** $pV = nRT \Rightarrow \dfrac{p_1V_1}{n_1T_1} = \dfrac{p_2V_2}{n_2V_2}$

We have $\quad T_1 = 273+40 = 313\ \text{K} \qquad p_1 = 101\ \text{kPa} + 608\ \text{kPa} = 709\ \text{kPa}$

$$n_2 = \tfrac{3}{4}n_1 \quad V_2 = V_1 \qquad T_2 = 273+315 = 588\ \text{K}$$

Then $P_2 = (709\ \text{kPa})\left(\frac{3}{4}\right)(588\ \text{K})/(313\ \text{K}) = 999\ \text{kPa}$. The final gauge pressure is 999 – 101 = 898 kPa.

**239.** Viewed in the center-of-mass reference frame, the molecules of a gas are in random motion, with a wide distribution of kinetic energies. The *root-mean-square speed* (or *thermal speed*) $v_{\text{rms}}$ may be defined as the speed of a hypothetical molecule whose translational kinetic energy equals the average translational kinetic energy over the whole ensemble of gas molecules: $\frac{1}{2}mv_{\text{rms}}^2 \equiv \overline{\frac{1}{2}mv^2}$, or $v_{\text{rms}} = \sqrt{\overline{v^2}}$. For a dilute gas, obey the ideal gas law, $\frac{1}{2}mv_{\text{rms}}^2 = \frac{3}{2}kT$. Thus, the absolute temperature of a macroscopic object is a measure of the average translational kinetic energy of its molecules, as determined in the object's center-of-mass frame. This last result can be combined with the ideal gas law, $pV = NkT$, to yield $pV = \frac{1}{3}Nmv_{\text{rms}}^2$. Noting that $Nm = M$, the total mass of the gas sample, we have $p = \frac{1}{3}pv_{\text{rms}}^2$.

**240.** The mass of an $H_2$ molecule may be calculated from the molecular weight as

$$m = \frac{M}{N_0} = \frac{2.016\times10^{-3}\ \text{kg/mol}}{6.02\times10^{23}\ \text{mol}^{-1}} = 3.35\times10^{-27}\ \text{kg}$$

Then $\dfrac{1}{2}mv_{\text{rms}}^2 = \dfrac{3}{2}kT \quad v_{\text{rms}} = \sqrt{\dfrac{3kT}{m}} = \sqrt{\dfrac{3(1.38\times10^{-23})(373)}{3.35\times10^{-27}}} = 2200\ \text{m/s} = 2.2\ \text{km/s}$

**241.** **(A)** Since $v \propto T^{1/2}$, $v = v_0(573/293)^{1/2} = 1.40v_0$.

**(B)** Since there is no pressure dependence at constant $T$, the speed is still $v_0$.

**(C)** Note that $mv^2$ is constant, so if $m$ is tripled, then $v^2$ will be one-third as large. The speed would be $(v_0/3)^{1/2} = 0.58v_0$.

**242.** Use the fact that $(m_0\langle v^2\rangle)/2 = (3kT)/2$. Then $\langle v^2\rangle = [3(1.38\times10^{-23})(3)]/(1.67\times10^{-27}) = 7.44\times10^4$; thus, $v_{rms} = 272$ m/s.

**243.** The average distance traveled by a gas molecule between collisions. For an ideal gas of spherical molecules with radius $b$,

$$\text{mean free path} = \frac{1}{4\pi\sqrt{2}b^2(N/V)} \tag{1}$$

where $N/V$ is the number of molecules per unit volume.

## Chapter 15: The First Law of Thermodynamics

**244.** Work done on a gas is considered positive; work done by a gas is considered negative. This means that when a gas is compressed (i.e., when its volume decreases), the variable $W$ in the first law of thermodynamics is positive. And, when the gas expands (i.e., when its volume increases), the variable $W$ is negative.

**245.** The internal energy ($U$) of a system is the total energy content of the system. It is the sum of the kinetic, potential, chemical, electric, nuclear, and all other forms of energy possessed by the atoms and molecules of the system. A form of energy may be categorized as organized or disorganized. *Organized energy* is associated with concerted behavior of the particles composing the system, e.g., macroscopic motion of the system. Also, chemical potential energy, where a definite amount of energy is to be released for each molecule formed, represents organized energy. *Disorganized energy* is associated with the random interactions ("collisions") of the particles. Thus, the temperature of an ideal gas measures its disorganized kinetic energy. Also called *thermal energy*, this category is of central interest in thermodynamics.

**246.** An isothermal process is one in which the system changes in such a way that the temperature remains constant throughout. It makes sense to talk about an isothermal process only for quasistatic processes in which there is a meaningful system temperature at all times, and the system thus progresses in orderly fashion from one state to another.

An isobaric process is one in which the pressure on the system remains unchanged throughout the process. It applies primarily to quasistatic processes in which a definite system pressure (or pressure distribution) exists.

An isovolumic process, sometimes called isometric or isochoric, is one in which the volume of the system remains the same. It is meaningful in both quasistatic and more violent processes.

An adiabatic process is one in which no heat transfer takes place into or out of the system. It is meaningful in both quasistatic and more violent processes.

Almost all questions in this chapter refer to quasistatic processes.

**247.**

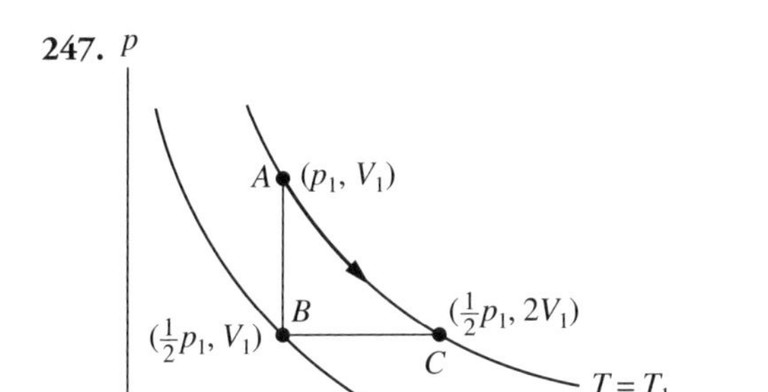


**Figure A15.1**

**(A)** Refer to Figure A15.1; $W_{AB} = 0$, since the volume does not change.
**(B)** By the ideal gas law at constant volume,

$$\frac{T_2}{T_1} = \frac{p_2}{p_1} = \frac{1}{2} \qquad \text{or} \qquad T_2 = \frac{T_1}{2} = 137\ \text{K}$$

**(C)** The constant-pressure process returns the gas to the original temperature, $T_1 = 273$ K, since pressure is constant, and $V$ doubles. $W_{BC}$ is the area under the graph from B to C, equal to ½ $P_1V_1$.

$$W_{BC} = \int_{V_1}^{2V_1} \frac{1}{2} p_1\ dV = \frac{1}{2} p_1 V_1 = \frac{1}{2} RT_1 = \frac{1}{2}(8.31)(273) = 1140\ \text{J}$$

$$W_{BC} = \text{area under graph}\left(\frac{1}{2} P_1\right)(2v_1 - v_1)$$

$$= \frac{1}{2} P_1 V_1 = \frac{1}{2} \pi RT$$

$$= \frac{1}{2}(1001)\left(8.3 \frac{\text{J}}{\text{mol} \cdot \text{K}}\right)(273\ \text{K}) = 1100\ \text{J}$$

**248.** **(A)** Internal energy for an ideal gas is linearly related to the temperature of the gas with $U = \frac{3}{2}\pi RT$, so the answer to parts **(A)** and **(B)** is the same: $\Delta U = \frac{3}{2}\pi R \Delta T = \frac{3}{2}(100\ \text{mol})(8.3\ \frac{\text{J}}{\text{mol}\cdot\text{K}})(20°\text{C}) = 42$ J.

**249.** **(A)** In this case, $Q = 0$, so $\Delta U = -W = -(-45\ \text{J}) = 45$ J.
**(B)** The heat flow in the adiabatic process is zero.

**250.** **(A)** The work done by the water is given by

$$W = p_{\text{atm}} \Delta V = (1.013 \times 10^5\ \text{N/m}^2)[1.671 - 0.001)\ \text{m}^3] = 169\ \text{kJ}$$

**(B)** The latent heat of vaporization of water is $L = 540$ kcal/kg. This is the heat per kilogram that must be added to vaporize 100-°C water at a constant pressure of 1 atm.

The change in the internal energy of 1 kg of water when it is boiled at 1 atm is therefore given by

$$\Delta U = Q - W = Lm - W = (540 \text{ kcal/kg})(1.00 \text{ kg})(4.184 \text{ kJ/kcal}) - (169 \text{ kJ})$$
$$= 2090 \text{ kJ}$$

**(C)** $\Delta U = Q + W$

$Q = -5000$ J (heat removed, so −)

$W = 0$ (no volume change, so $W = 0$)

$\Delta U = -5000$ J

**251.** Applying the gas law, we have

$$T_A = \frac{p_A V_A}{nR} = \frac{(1.013\times10^5 \text{ Pa})(22.4 \text{ m}^3)}{(1.00 \text{ kmol})(8.314\times10^3 \text{ J/kmol}\cdot\text{K})} = 273 \text{ K}$$

Because the process *AB* shown in Figure 251.1 is isometric, $T_B = (p_B/p_A)T_A$. With $p_B =$ 2.00 atm $= 2p_A$, we find that $T_B = 546$ K. The process *CA* is isobaric, so Charles' law applies. Thus, $V_C/T_C = V_A/T_A$, so that $V_C = (T_C/T_A)V_A$. But *BC* is an isothermal process, so that $T_C = T_B = 2T_A$. Therefore, $V_C = 2V_A = 44.8 \text{ m}^3$.

**252.**

| Path | *Q* | *W* | *U* | *T* |
|---|---|---|---|---|
| *KL* | + | + | + | + |
| *LM* | – | 0 | – | – |
| *MN* | – | – | 0 | 0 |
| *NK* | 0 | – | + | + |

**Figure A15.2**

Path *KL:* The process is an isobaric expansion ($p$ remains constant while $V$ increases). Clearly, the gas does work on its surroundings, so $W > 0$. For the pressure to remain constant while the gas density decreases, the temperature must increase: $\Delta T > 0$. For an ideal gas, the energy is a function of only the temperature, and $U$ increases with $T$, so $\Delta U > 0$. Because $Q = \Delta U + \Delta W$, we must have $Q > 0$.

Path *LM:* The process is isometric cooling ($V$ remains constant while $p$—and with it $T$—decreases). Because the volume does not vary anywhere along the path, we have $\Delta V \equiv 0$ and therefore $W = 0$. As we already noted, $\Delta T < 0$, since that is the only way that $p$ can decrease at constant volume. Since $U$ is a monotonically increasing function of $T$, we have $\Delta U < 0$. Then $\Delta U W = Q < 0$.

Path *MN:* The process is isothermal compression, so $\Delta T = 0$. Therefore, the ideal gas has $\Delta U = 0$. There is a monotonic volume decrease ($\Delta V < 0$), so the work $= W < 0$. Therefore, $\Delta U + \Delta W = \Delta Q < 0$.

Path *NK:* The process is adiabatic compression, so that $\Delta Q = 0$. There is a monotonic volume decrease ($\Delta V < 0$), so $W < 0$. Therefore, $\Delta Q - W \equiv \Delta U > 0$. The temperature of the ideal gas must increase correspondingly: $\Delta T > 0$.

These results are summarized in Figure A15.2.

**253.** In expansion, the work done is equal to the area under the pertinent portion of the $p$-$V$ curve but is negative. In contraction, the work is numerically equal to the area but is positive.

**(A)** work = area $ABFEA = [(4-1.5)\times 10^{-6}\ \text{m}^3](4\times 10^5\ \text{N/m}^2) = 1.00\ \text{J}$

**(B)** work = area under $BC = 0$

In portion $BC$, the volume does not change; therefore, $p\,\Delta V = 0$.

**(C)** This is contraction and so the work is positive.

$$\text{work} = -(\text{area } CDEFC) = -(2.5\times 10^{-6}\,\text{m}^3)(2\times 10^5\ \text{N/m}^2) = +0.50\ \text{J}$$

**(D)** work = 0

**254. (A)** Method 1: From Question 253, the net work done is 1.00 J – 0.50 J = 0.50 J. Method 2: The net work done is equal to the area enclosed by the $p - V$ diagram.

$$\text{work} = \text{area } ABCDA = (2\times 10^5\ \text{N/m}^2)(2.5\times 10^{-6}\,\text{m}^3) = 0.50\ \text{J}$$

**(B)** Suppose that the cycle starts at point $A$. The gas returns to this point at the end of the cycle, so there is no difference in the gas at its start and endpoint. For one complete cycle, $\Delta U$ is therefore zero. We have then, if the first law is applied to a complete cycle,

$$Q = \Delta U - W = 0 - (-50\ \text{J})\ 0.50\ \text{J} = +0.50\ \text{J}$$

**255.** During the heating process, the internal energy changed by $\Delta U_1$ and work $W_1$ was done. The gas pressure was

$$p = \frac{8(9.8)\ \text{N}}{60\times 10^{-4}\ \text{m}^2} + 1.00\times 10^5\ \text{N/m}^2 = 1.13\times 10^5\ \text{N/m}^2.$$

## Chapter 16: The Second Law of Thermodynamics

**256.** In an idealized *reversible process*, a system goes from an initial equilibrium state to a final equilibrium state through a continuous sequence of equilibrium states. This means that at every instant during the process the system is in thermal and mechanical equilibrium with its surroundings. The direction of such a process can be reversed (hence, “reversible”) at any instant by an infinitesimal change in external conditions.

An actual process approaches reversibility to the degree that it is quasistatic (i.e., extremely slow) and that dissipative effects (e.g., friction) are absent.

**257.** A *heat engine* is a device or system that converts heat into work. Heat engines operate by absorbing heat from a reservoir at a high temperature, performing work, and giving off heat to a reservoir at a lower temperature. The *efficiency* $\eta$ of a cyclic heat engine is

$$\eta = \frac{W}{Q_{\text{hot}}} = 1 - \frac{Q_{\text{cold}}}{Q_{\text{hot}}}$$

where $Q_{\text{hot}}$, $Q_{\text{cold}}$, and $W$ (see Figure A16.1) represent, respectively, the heat absorbed per cycle from the higher-temperature reservoir, the heat rejected per cycle to the lower-temperature reservoir, and the work carried out per cycle. The second expression above for $\eta$ follows from $W = Q_{hot} - Q_{cold}$, since $\Delta U = 0$ over a cycle.

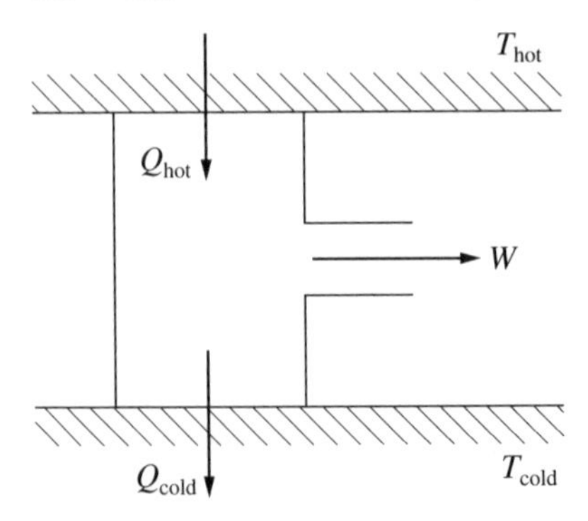


**Figure A16.1**

The greatest possible thermal efficiency of an engine operating between two heat reservoirs is that of a *carnot engine*, one that operates in the Carnot cycle. This maximal efficiency is

$$\eta_{max} = \frac{T_{hot} - T_{cold}}{T_{hot}}$$

**258.** $$\text{maximum efficiency} = \frac{T_h - T_c}{T_h} = \frac{480 - 300}{480} = 37.5\%$$

$$\text{actual efficiency} = \frac{\text{energy output}}{\text{energy input}} = \frac{1.2}{1 \times 4.184} = 28.7\%$$

The actual efficiency is about three-fourths of the maximum.

**259.** $$\text{efficiency} = \frac{Q_h - Q_c}{Q_h} = \frac{T_h - T_c}{T_h} \qquad Q_h = (1\text{ kcal})(4.184\text{ kJ/kcal}) = 4.184\text{ kJ}$$

$$\text{efficiency} = \frac{Q_h - Q_c}{4.184\text{ kJ}} = \frac{700\text{ K} - 450\text{ K}}{700\text{ K}} \qquad \text{Then} \qquad W = Q_h - Q_c = 1.49\text{ kJ}$$

**260.** $$\text{efficiency} = \frac{Q_h - Q_c}{Q_h} = \frac{T_h - T_c}{T_h} \qquad \frac{500 - Q_c}{500} = \frac{590 - 390}{590}$$

$$500 - Q_c = 169\text{ kcal} \qquad \text{and} \qquad Q_c = 331\text{ kcal delivered to the sink}$$

$W = Q_h - Q_c = 169\text{ kcal} = 169(4.184) = 710\text{ kJ}$.

**261.** The first law tells us that $Q = \Delta U + W$. Because the process was isothermal, the internal energy of the ideal gas did not change. Therefore, $\Delta U = 0$ and $Q = W = -730$ J.

(Because the gas was compressed, the gas did negative work—hence the minus sign.) Now we can write

$$\Delta S = \frac{Q}{T} = \frac{-730 \text{ J}}{293 \text{ K}} = -2.5 \text{ J/K}$$

Note that the entropy change is negative. Disorder of the gas decreased as it was pushed into a smaller volume.

**262.** The second law statement refers to the change in entropy *of the universe*. Clearly, the heat that left the cylinder entered some other system in the environment, causing an increase in entropy in that system. The second law says that this increase was at least as great as the decrease in entropy of the gas in the cylinder.

## Chapter 17: Wave Motion

**263.** It took sound 6.0 s to travel from the flash to the observer. Use $x = vt = (330 \text{ m/s}) \times (6.0 \text{ s}) = 2000 \text{ m} = 2 \text{ km}$.

**264.** $\lambda = \dfrac{v}{f} = \dfrac{3\times 10^8 \text{ m/s}}{760{,}000 \text{ Hz}} = 395 \text{ m}$

**265.** $v = \lambda f \quad 1530 = 1800\lambda \quad \lambda = 0.85 \text{ m}$

**266.** The frequency of the disturbance, $v$, stays the same.

$$v = \frac{u}{\lambda} = \frac{4v}{\lambda_2} \qquad \text{whence } \lambda_2 = 4\lambda$$

**267.** **(A)** $v = \lambda f \qquad 5050 = 40{,}000\lambda \qquad \lambda = 0.13 \text{ ft}$

**(B)** If the submarine is at a distance $d$, the sonar wave must travel a total distance $2d$ from the transducer to the submarine and, after reflection, back to the receiver.

$$v = \frac{2d}{t} \qquad 5050 = \frac{2d}{5.0} \qquad d = 12{,}600 \text{ ft}$$

**268.** **(A)** $v = \lambda f = (0.31 \text{ m})(1.20 \text{ Hz}) = 37 \text{ m/s}$

**(B)** $v = \sqrt{S/\mu}$, where $S$ is the tension. Then $\mu = S/v^2 = (1.20 \text{ N})/(37 \text{ m/s})^2 = 8.76\times 10^{-4} \text{ kg/m} \qquad M = \mu L = (0.50 \text{ m})(8.76\times 10^{-4} \text{ kg/m}) = 4.38\times 10^{-4} \text{ kg} = 0.44 \text{ g}$

**269.** $v \propto T^{1/2}$; thus, the velocity doubles and $v = 2000$ ft/s.

**270.** **(A)** Amplitude, $A = 3 \text{ mm} = 0.3 \text{ cm}$.

**(B)** $f$ is the same as the vibrator frequency, so $f = 60$ Hz.

**(C)** $\lambda$ = distance of one repeat of the wave $\lambda = 2$ cm.

**(D)** Speed, $v = \lambda f = (2.0 \text{ cm})(60 \text{ Hz}) = 120 \text{ cm/s}$.

**(E)** Period, $T = 1/f = 0.0167$ s.

**271.** By comparison with the standard form $y = y_0 \sin[2\pi f(t - x/v)]$, we find $y_0 = 0.02$ m, $f = 4.78$ Hz, $v = 7.5$ m/s; and since $\lambda = v/f$, $\lambda = 1.57$ m.

**272.** The amplitude = 0.06 m and $\frac{5}{2}\lambda = 20$ cm, so $\lambda = 0.080$ m; then $f = v/\lambda = 300/0.080 = 3750$ Hz. Now $(2\pi)/\lambda = 78.5$ and $2\pi f = 23600$, so $y = 0.06 \sin(78.5x - 23600t)$ m.

**273.** $y = (0.20 \text{ cm}) \sin(188t - 10.5x)$. For any fixed $x$, this is the equation for simple harmonic motion with amplitude $y_0 = 0.20$ cm and angular frequency $\omega = 188$ rad/s. Then $v_{y,\max} = \omega y_0 = 37.6$ cm/s; $a_{y,\max} = \omega^2 y_0 = 7070$ cm/s$^2$.

**274.** $v = \sqrt{\dfrac{T_s}{\mu}} = \sqrt{\dfrac{4}{0.0001}} = 200$ m/s

A standing wave is of the fundamental frequency if one half-wavelength occupies the length of the wire; i.e., if $\lambda = 1.0$ m. Thus

$$v = \frac{v}{\lambda} = \frac{200 \text{ m/s}}{1.0 \text{ m}} = 200 \text{ Hz}$$

**275.** $v = \sqrt{\dfrac{T_s}{\mu}} = \sqrt{\dfrac{100}{0.30}} = 18.3$ m/s $\quad s = vt \quad$ so $\quad 5.0 = 18.3t \quad$ and $\quad t = 0.27$ s

**276.** **(A)** $5\left(\dfrac{\lambda}{2}\right) = 12$ m $\quad$ or $\quad \lambda = 4.8$ m

**(B)** $v = \dfrac{20 \text{ m/s}}{4.8 \text{ m}} = 4.17$ Hz

**277.** The fundamental frequency $f_1$ corresponds to a wavelength of $2(0.4) = 0.8$ m; thus,

$$f_1 = \frac{418 \text{ m/s}}{0.8 \text{ m}} = 520 \text{ Hz}$$

The second and third harmonic frequencies are then $2f_1 = 1040$ Hz and $3f_1 = 1560$ Hz.

**278.** The third overtone is the fourth harmonic $f_4$. $f_4 = 1200$ Hz implies that $f_1 = 1200/4 = 300$ Hz. Then $f_2 = 2f_1 = 600$ Hz, and $f_3 = 3f_1 = 900$ Hz. $v = \lambda_4 f_4 = (1 \text{ m})(1200 \text{ Hz}) = 1200$ m/s [or $v = \lambda_1 f_1 = (4 \text{ m})(300 \text{ Hz}) = 1200$ m/s].

**279.** **(A)** The fundamental is $f_1 = v/(2L)$, the $n$th harmonic frequency is $nf_1 = 85$ Hz, while the next harmonic frequency is $(n+1)f_1 = 102$ Hz. From these we obtain $f_1 = 17$ Hz and $n = 5$.

**(B)** For $n = 5$, there are five segments; the length of each is $160/5 = 32$ cm.

**(C)** From $f_1 = 17 = v/[2(1.6)]$, $v = 54.4$ m/s.

**280.** In the third harmonic, $(3\lambda)/2 = 2.00$ m so that $v = f\lambda = 480(1.33) = 640$ m/s. We then use the relation $v^2 = T/\mu$ to find $\mu = (mg)/v^2 = [0.800(9.80)]/640^2 = 1.91 \times 10^{-5}$ kg/m.

**281.** As can be seen from Figure A17.1, $n(\lambda/2) = L$ holds for an open organ pipe as well as a string. Thus, the fundamental frequency is $f_1 = v/2l = (340\,\text{m/s})/2(30\text{ m}) = 570$ Hz. The first overtone is the second harmonic, and $v_2 = 2f_1 = 2(570) = 1140$ Hz. Similarly, $f_3 = 3f_1 = 3\,(570) = 1710$ Hz (second overtone).

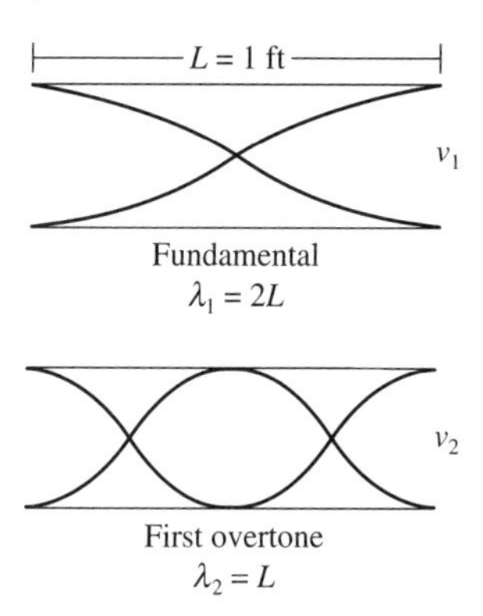


**Figure A17.1**

**282.** Closed pipe: $f_1 = \dfrac{V}{4L}$, and overtones must be odd multiples of the fundamental $f_1$.

$$f_1 = \frac{340\ \frac{\text{m}}{\text{s}}}{4(.77\text{ m})} = 110\text{ Hz}$$

The next two overtones are $3f_1$ and $5f$,

so 330 Hz and 550 Hz

**283.** As Figure A17.2 shows, the distance between water levels is the distance between successive nodes, or half a wavelength.

$$\frac{\lambda}{2} = 73 - 20 = 53\text{ cm} \qquad \text{and} \qquad \lambda = 106\text{ cm} = 1.06\text{ m}$$

$$v = f\lambda = 320 \times 1.06 = 339\text{ m/s}$$

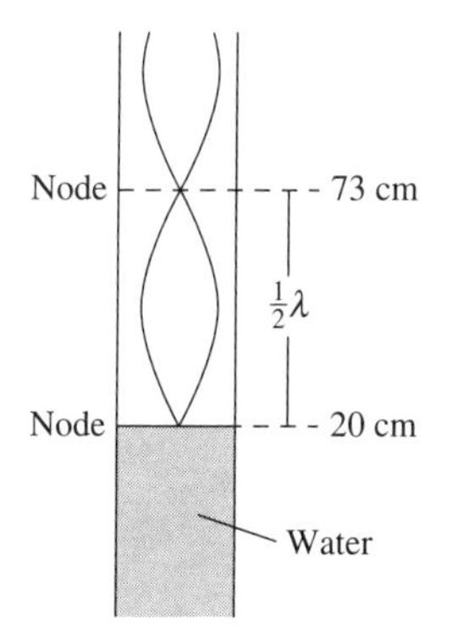


**Figure A17.2**

**284.** The brass rod is "open" at both ends, so the longitudinal wave will have $L = n(\lambda/2) = n[v/(2f)]$. For the fundamental, $0.40 = (v/6000)$, from which $v = 2400$ m/s.

## Chapter 18: Sound

**285.** The number of complete waves passing any point in air and in water in unit time is the same, so $f = 1000$ Hz for both media. Therefore,

$$\lambda_w = \frac{v_w}{f} = \frac{1500 \text{ m/s}}{1000 \text{ s}^{-1}} = 1.5 \text{ m}$$

**286.** As indicated in Problem 285, $f$ remains constant—in this case, at 60 kHz. The wavelength $= v/f = 330/(6.0 \times 10^4) = 5.5$ mm.

**287.** Each pipe is occupied by a quarter-wave (refer to Question 282); hence,

$$5 \text{ Hz} = f_1 - f_2 = \frac{340 \text{ m/s}}{4.4 \text{ m}} - \frac{340 \text{ m/s}}{4L}$$

Solving, $L = 1.18$ m.

**288.** Each pipe is occupied by a half-wave (see Question 281), so

$$f_1 - f_2 = \frac{1100 \text{ ft/s}}{4.8 \text{ ft}} - \frac{1100 \text{ ft/s}}{5.0 \text{ ft}} = 9 \text{ Hz}$$

**289.** The frequency of the second fork must be higher than that of the first fork, or adding the tape would have *increased* the number of beats. Therefore, $f_2 - 180 = 4$ or $f_2 = 184$ Hz.

**290.** Use the Doppler-effect equation:

$$\frac{f_L}{v + v_L} = \frac{f_s}{v - v_s}$$

where $v_L$ = velocity of listener relative to medium toward source, and $v_s$ = velocity of source relative to medium toward listener, $v$ = speed of sound in medium. Since the listener is not moving, $v_L$ is zero.

$$\frac{f_L}{1100} = \frac{400}{1100 - 100} \qquad f_L = \frac{400(1100)}{1000} = 440 \text{ Hz}$$

**291.** $\dfrac{f_L}{v + v_L} = \dfrac{f_s}{v - v_s} \qquad v_s = 0 \qquad$ Then

$$\frac{400}{1080 + 90} = \frac{f_s}{1080} \qquad \text{and} \qquad f_s = \frac{400(1080)}{1170} = 369 \text{ Hz}$$

**292. (A)** We let $v$ represent the emitted frequency, $|v|$ represent the sound speed, and $|v_s|$ represent the source speed. Since the bird is flying directly away from the observer, the received frequency $v'$ is given by

$$f' = f\frac{|v|}{|v| + |v_s|}$$

Adopting $|v| = 340$ m/s, and using the given values $|v_s| = 15$ m/s and $v = 800$ Hz, we find

$$f' = \frac{(800)(340)}{340+15} = 766 \text{ Hz}$$

**(B)** Whatever frequency is incident upon the cliff is reflected without change. Therefore, the observer will receive the same frequency in the echo that another observer on the cliff would hear directly. The frequency v″ in the echo is therefore given by

$$f'' = f\frac{|v|}{|v|-|v_s|} = \frac{(800)(340)}{(340-15)} = 837 \text{ Hz}$$

## Chapter 19: Coulomb's Law and Electric Fields

**293.** Use Coulomb's law to find $\mathbf{F}_1$ and $\mathbf{F}_2$ and take the vector sum.

$$F_1 = k\frac{Q_1Q}{r^2} = (9\times10^9)\frac{(6\times10^{-6})(2\times10^{-6})}{0.05^2} = 43.2 \text{ N} \quad \text{away from } Q_1$$

$$F_2 = (9\times10^9)\frac{(4\times10^{-6})(2\times10^{-6})}{0.05^2} = 28.8 \text{ N} \quad \text{toward } Q_1$$

$$F = F_1 - F_2 = 14.4 \text{ N} \quad \text{away from } Q_1$$

**294.** $F = F_1 + F_2 \quad F_1 = \frac{kQ_1Q_3}{r^2} = \frac{(9\times10^9)(20\times10^{-6})^2}{4^2} = 0.225 \text{ N}$

$$F_2 = \frac{kQ_2Q_3}{r^2} = \frac{(9\times10^9)(20\times10^{-6})^2}{2^2} = 0.9 \text{ N} \qquad F = 0.225 + 0.9 = 1.1 \text{ N} \quad \text{to the right}$$

**295.**

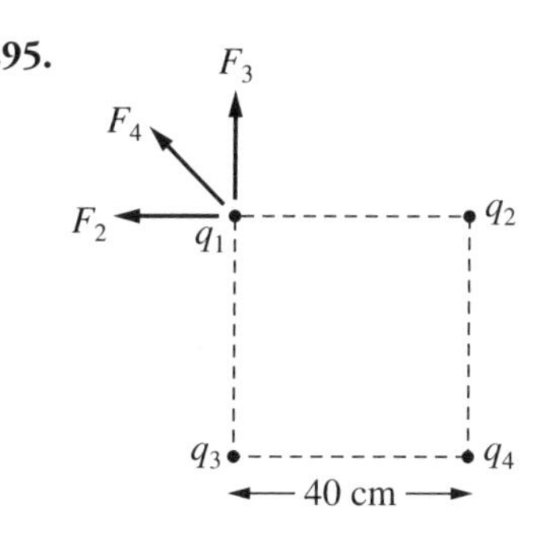


**Figure A19.1**

The situation is depicted in Figure A19.1. We consider the forces acting on $q_1$, depicted in the diagram by $F_2$, $F_3$, $F_4$, where the labels identify the respective charges exerting the force. By symmetry $F_2 = F_3 = [(9\times10^9)(3\times10^{-6})^2]/0.40^2$, or $F_2 = F_3 = 0.51$ N. Since the directions of these forces are along the edges, as shown, their vector sum will lie along the diagonal from $q_4$ to $q_1$ and have magnitude $F_2\cos45° + F_3\cos45° = [2(0.51)]/\sqrt{2} = 0.72$ N. The remaining force $F_4$ is also along this diagonal, and $F_4 = [(9\times10^9)(3\times10^{-6})^2]/(0.40\times\sqrt{2})^2 = 0.25$ N. The resultant of all three forces thus points along the diagonal and away from the square and has magnitude 0.97 N.

**296.**

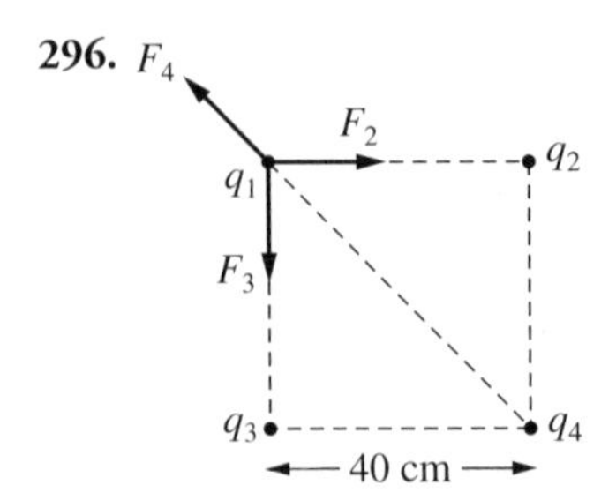


**Figure A19.2**

The new situation (Figure A19.2) is identical to that of Question 295, except that now $q_1 = q_4 = -3\ \mu C$. Again we calculate the resultant force on $q_1$. $F_2$ and $F_3$ have the same magnitudes as in Problem 295, but the directions are opposite, as shown. The vector sum of $F_2$ and $F_3$ is now 0.72 N pointed inward along the diagonal. $F_4$ is again repulsive and, as before, is 0.25 N. The resultant of all three forces is now $0.72\ N - 0.25\ N = 0.47\ N$ pointed inward along the diagonal (toward $q_4$).

**297.**

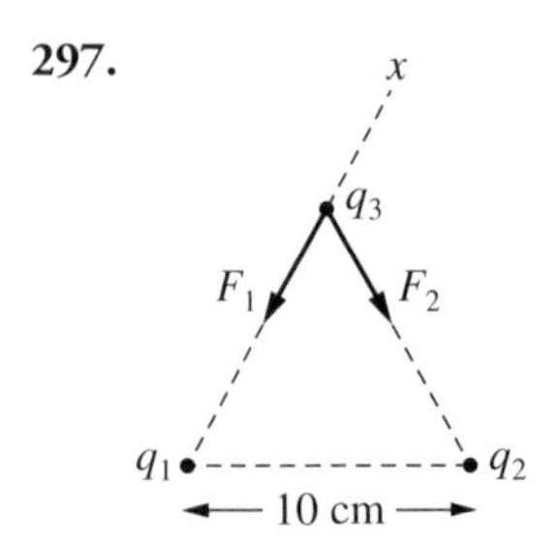


**Figure A19.3**

The situation is as shown in Figure A19.3 with $q_1 = 2\ \mu C$, $q_2 = 3\ \mu C$, and $q_3 = -8\ \mu C$. For definiteness, we let the side from $q_1$ to $q_3$ be our $x$ axis. The net force on $q_3$ is the vector sum of $F_1$ and $F_2$, the forces due to $q_1$ and $q_2$. As shown, these are attractive forces. In magnitude, $F_1 = [(9\times10^9)(2\times10^{-6}C)(8\times10^{-6}C)]/(0.10\ m)^2$, or $F_1 = 14.4$ N. Similarly, $F_2 = [(9\times10^9)(3\times10^{-6})\,(8\times10^{-6})]/0.10^2 = 21.6$ N. Let $\mathbf{F} = \mathbf{F}_1 + \mathbf{F}_2$; then $F_x = F_{1x} + F_{2x} = -14.4\ N - 21.6\ N\cos 60° = -25.2$ N. $F_y = F_{1y} + F_{2y} = 0 + 21.6\ N\sin 60° = 18.7$ N. $F = \sqrt{F_x^2 + F_y^2} = \sqrt{(25.2)^2 + (18.7)^2} = 31.4$ N.

**298.**

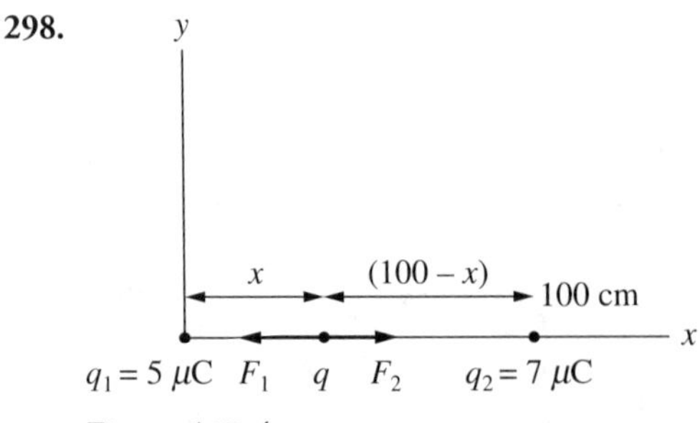


**Figure A19.4**

We assume that the appropriate location for our test charge is at position $x$, as shown in Figure A19.4. If the test charge $q$ is negative, $q_1$ and $q_2$ will attract $q$ in opposite directions; if it is positive, $q_1$ and $q_2$ will repel $q$ in opposite directions. In either case, the condition for $q$ to experience no net force is that $F_1 = F_2$, or $(kqq_1)/x^2 = (kqq_2)/(1.0 - x)^2$, where we convert distance to meters. Dividing out $k$ and $q$, we have $q_1/x^2 = q_2/(1.0 - x)^2$. (Note that our result, as might be expected, does not depend on $q$.) Substituting in numbers and cross-multiplying, we get $5(1 - x)^2 = 7x^2$, or $2x^2 + 10x - 5 = 0$. Solving $x = (-10 \pm \sqrt{100 + 40})/4 = \{0.46 \text{ m}, -5.46 \text{ m}\}$. Since $x$ for our case must lie between the two charges, our answer is 46 cm. (At $x = -5.46$ m, the two forces are equal in magnitude *and* direction.)

**299. (A)** At first the force between them is *attractive* with magnitude $F = [(9 \times 10^9) \times (3 \times 10^{-9})](12 \times 10^{-9})]/0.03^2 = 3.6 \times 10^{-4}$ N.

**(B)** Since the two balls are metallic and identical, when they are touched the charges rearrange themselves to a new equilibrium distribution that must have a like charge on each ball. Since the total available charge is −9 nC, each ball has −4.5 nC of charge. When they are moved 3 cm apart, the new force is *repulsive* and has magnitude $F = [(9 \times 10^9)(4.5 \times 10^{-9})(4.5 \times 10^{-9})]/0.03^2 = 2.0 \times 10^{-4}$ N.

**300.** Since the system is at rest, we can apply the conditions for equilibrium to the ball on the left. Note that three forces act on the ball: its weight $mg$; the tension $T$ in the string; and $F$, the repulsive force due to the charge on the other ball. We have the usual conditions for equilibrium: $\Sigma F_x = 0$, from which $F - 0.6T = 0$ and $\Sigma F_y = 0$, which gives $0.8T - (0.2 \times 10^{-3}\text{kg})(9.8 \text{ m/s}^2) = 0$, or $T = 2.45 \times 10^{-3}$N. Using this to find $F$, we obtain $F = 1.47 \times 10^{-3}$N. This is the force obeying Coulomb's law.

Substituting in Coulomb's law, we have

$$1.47 \times 10^{-3} = (9 \times 10^9)\frac{q^2}{(0.60)^2}$$

where all units are SI units. Solving for $q$, we find that $q \approx 2.4 \times 10^{-7}\,\text{C} = 0.24\ \mu C$.

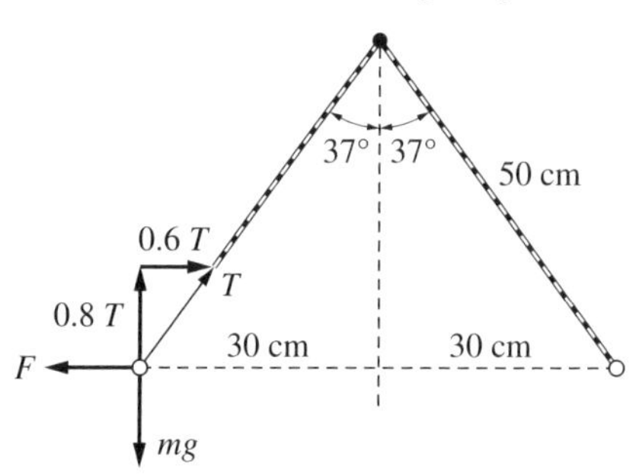


**Figure A19.5**

**301.** From Coulomb's law,

$$F = \frac{1}{4\pi e_0}\frac{|q||q|}{r^2} = (8.99 \times 10^9\,\text{N} \cdot \text{m}^2/\text{C}^2)\frac{(1.60 \times 10^{-19}\,\text{C})^2}{(5.29 \times 10^{-11}\,\text{m})^2} = 82.3 \text{ nN}$$

$$a = \frac{F}{m} = \frac{82.3 \times 10^{-9}\,\text{N}}{9.11 \times 10^{-31}\,\text{kg}} = 9.03 \times 10^{22}\,\text{m/s}^2$$

The orbital speed is given from expression $a = v^2/r$ for centripetal acceleration; thus, it is

$$v = (ar)^{1/2} = (9.03\times10^{22}\ \text{m/s}^2 \times 5.29\times10^{-11}\ \text{m})^{1/2} = 2.19\times10^6\ \text{m/s}$$

Since this speed is less than 1 percent of the speed of light, the nonrelativistic form of Newton's second law, $a = F/m$, was appropriate.

**302.** As always, assume that the charges giving rise to the electric field are stationary and not affected by the charge brought into the field.

**(A)** Since $\mathbf{F} = q\mathbf{E}$, we must have $\mathbf{E}$ in the $+x$ direction with magnitude $E = F/q = (2\times10^{-3}\ \text{N})/(6\times10^{-6}\ \text{C}) = 333\ \text{N/C}$.

**(B)** $\mathbf{F}' = q'\mathbf{E}$ and since $q'$ is negative, $\mathbf{F}'$ is along the $-x$ direction. In magnitude, $F' = (2\times10^{-6}\ \text{C})(333\ \text{N/C}) = 0.67\ \text{mN}$.

**303.** The fields due to these point charges obey the superposition principle. The fields due to each charge are shown in the figure. Using $E = (kq)/r^2$ for each charge, we get

$$E_2 = 72\ \text{kN/C} \qquad E_8 = 288\ \text{kN/C} \qquad E_{12} = 432\ \text{kN/C}$$

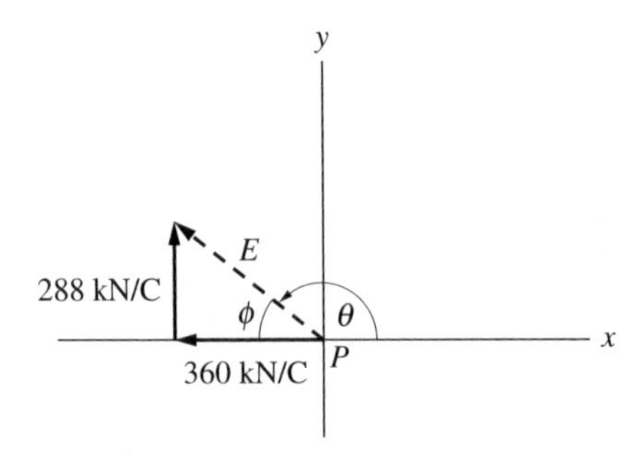


**Figure A19.6**

We therefore find at $P$ that $E_x = -360$ kN/C and $E_y = 288$ kN/C (see Figure A19.6). From this, $\mathbf{E} = 461$ kN/C at $\theta = 141°$.

**304.**

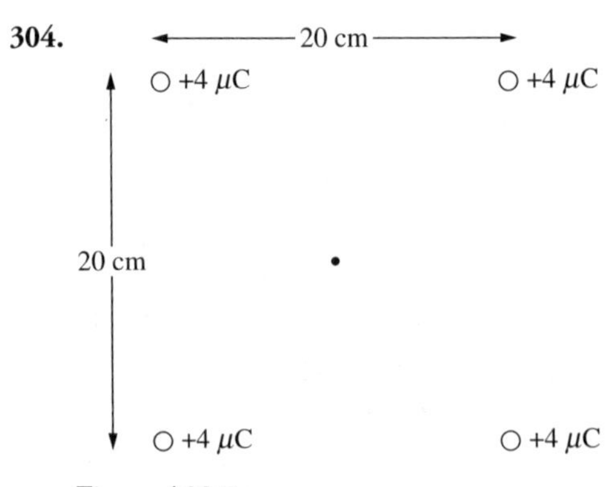


**Figure A19.7**

$\vec{E}$ for each charge has magnitude $\frac{kQ}{d^2}$, in a direction away from the charge.

The magnitude of $\vec{E}$ due to each charge is identical, since they are equal charges equidistant from the center point. The top-left charge produces a field opposite in direction to that of the bottom-right charge; these $\vec{E}$ vectors add to O. Similarly with the other two charges. Thus, the net electric field at the center is O.

**305.** $q_1 = q_4 = 4\mu C$; $q_2 = q_3 = -4\ \mu C$. Again, at the center, $E_1 = -E_4$, $E_2 = -E_3$, *so* $E = 0$.

**306.** Draw a vector diagram showing the fields produced at $A$ by each of the three charges. Then find the vector sum $E_t$.

$$E = \frac{kQ}{r^2} = \frac{(9\times10^9)(4\times10^{-6})}{(0.20)^2} = 0.9\ \text{MN/C}$$

$$E^+ = \sqrt{(9\times10^5)^2 + (9\times10^5)^2} = \sqrt{162\times10^{10}} = 1.27\ \text{MN/C}$$

$$E^- = \frac{(9\times10^9)(-4\times10^{-6})}{(0.283)^2} = -0.45\ \text{MN/C}$$

$$E_t = E^+ + E^- = 0.82\ \text{MN/C away from the negative charge.}$$

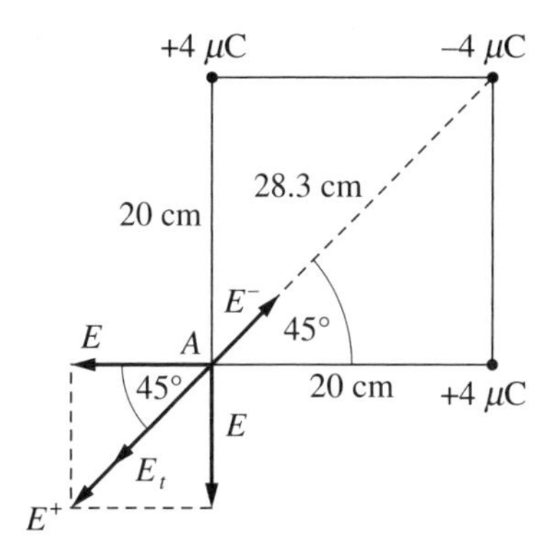


**Figure A19.8**

**307.** The ball is presumably in equilibrium under the action of three forces: the tension in the thread, $T$; the pull of gravity on the ball, $w = mg = 5.9 \times 10^{-3}$ N; and the coulomb force, $F_c$, which is downward if the charge, $q$, on the ball is positive, as shown in Figure A19.9, and upward if $q$ is negative.

**(A)** $F_c = qE = (8\times10^{-6}\text{C})(300\ \text{N/C}) = 2.4\times10^{-3}\text{N}$ downward. Then $T - w - F_c = 0$, or $T = 5.9\times10^{-3}\text{N} + 2.4\times10^{-3}\text{N} = 8.3\times10^{-3}\text{N}$.

**(B)** $F_c = |(-8\times10^{-6}\text{C})|\ (300\ \text{N/C}) = 2.4\times10^{-3}\text{N}$ in magnitude, but points upward. Then $T + F_c - w = 0$, or $T = w - F_c = 5.9\times10^{-3}\text{N} - 2.4\times10^{-3}\text{N}$, $T = 3.5\times10^{-3}\text{N}$.

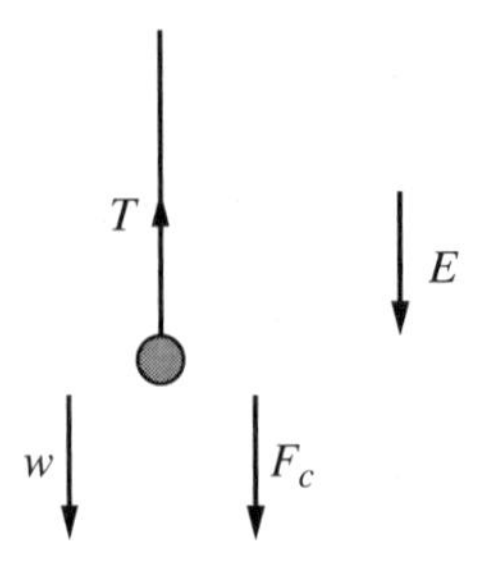


**Figure A19.9**

**308.** Let $w = mg = (0.60 \times 10^{-3}\ \text{kg})(9.8\ \text{m/s}^2) = 5.9$ mN be the weight of the ball. Let $F_c$ be the coulomb force due to the charge $q$ in the given electric field. Clearly, $F_c$ is to the left, so $q$ is negative. From equilibrium, we have $T\cos 20° = w = 5.9$ mN, and $T\sin 20° = F_c$. Solving, $T =$ 6.3 mN and so $F_c = 2.1$ mN. Finally $F_c = |q|\,E$, or $|q| = (2.1 \times 10^{-3}\text{N})/(700\ \text{N/C}) = 3.1\ \mu\text{C}$. Thus, $q = -3.1\ \mu\text{C}$.

**309.** The force, and hence acceleration, are constant and must be in the negative $x$ direction. This is necessary because if there were a $y$ component of force, the velocity component in that direction would continually increase and never become zero for $t > 0$. Furthermore, the component in the $x$ direction must act to slow down the initial velocity of the ball. From 1-D kinematics, $v_x^2 = v_{0x}^2 + 2a_x x$. For our case $v_{0x} = 3 \times 10^6\,\text{m/s}$, $v_x = 0$ and $x = 0.45$ m. Then $a_x = -(3 \times 10^6)^2/(2 \times 0.45) = -1.0 \times 10^{13}\,\text{m/s}^2$. The force causing this acceleration is $F_x = ma_x = (9.1 \times 10^{-31}\ \text{kg})(-1.0 \times 10^{13}\ \text{m/s}^2)$, or $F_x = -9.1 \times 10^{-18}$ N. $F_x$ is due to the electrostatic force on the electron caused by the electric field, so $F_x = qE_x$. $E_x = (-9.1 \times 10^{-18}\text{N})/(-1.6 \times 10^{-19}\text{C}) = 57$ N/C. Finally, since $F_y = 0$, we have the electric field = 57 N/C in the $+x$ direction. (*Note:* The force of gravity on the electron is negligible in comparison with the electric force and has therefore been ignored.)

**310.** This is a projectile problem, and the path followed by the particle will be as shown. Taking upward as positive and using $e = 1.6 \times 10^{-19}\,\text{C}$, $m = 9.1 \times 10^{-31}\,\text{kg}$, and $E = 10^3\text{N/C}$, we have for the vertical problem that

$$v_{0y} = 0 \qquad a = \frac{eE}{m} = 1.76 \times 10^{14}\,\text{m/s}^2 \qquad y = 0.5 \times 10^{-2}\,\text{m}$$

**311.** When the drop falls at its terminal velocity, the air friction force equals $mg$. To make it move upward with the same speed, one has to overcome gravity as well as air friction in the downward direction, so it takes a force of 2 mg upward. Each electron carries $1.60 \times 10^{-19}$ C of negative charge.

$$F = Eq = 2mg \qquad E(6 \times 1.60 \times 10^{-19}) = 2(1.6 \times 10^{-15})(9.8)$$

$$E = \frac{2 \times 10^{-15}}{6 \times 10^{-19}} 9.8 = 32.7\ \text{kN/C}$$

## Chapter 20: Electric Potential and Capacitance

**312.**

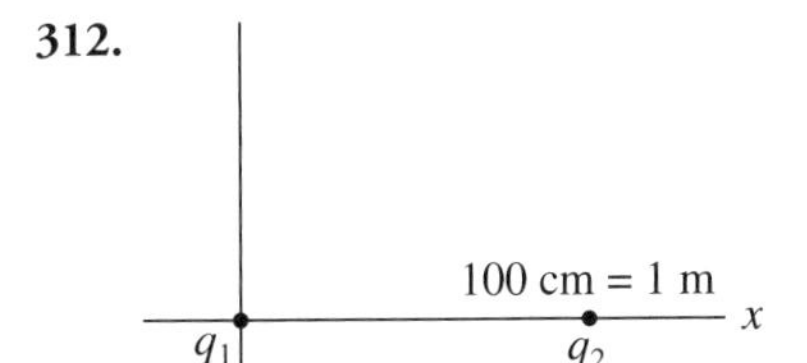


**Figure A20.1**

The two charges are as shown in Figure A20.1. At a point $x$ along the $x$ axis, the potential is

$$V(x) = k\left(\frac{q_1}{|x|} + \frac{q_2}{|x-1|}\right)$$

Setting $V = 0$ yields

$$\frac{2}{|x|} - \frac{3}{|x-1|} = 0 \qquad \text{or} \qquad 2|x-1| = 3|x|$$

We consider three cases: $x > 1$, $0 < x < 1$, and $x < 0$.

For $x > 1$: $2(x - 1) = 3x \Rightarrow x = -2$ (contradiction and no solution)
For $0 < x < 1$: $2(1 - x) = 3x \Rightarrow x = 0.4$ or $x = 40$ cm
For $x < 0$: $2(1 - x) = -3x \Rightarrow x = -2$ or $x = -200$ cm

**313.** $V_P = (9\times10^9)\left(\frac{6\times10^{-9}}{0.06} + \frac{6\times10^{-9}}{0.06} + \frac{6\times10^{-9}}{0.104}\right) = 2320 \text{ V}$

**314.** $E$ is constant so

$$E = \frac{V}{d} \quad \text{or} \quad 5000 = \frac{150}{d} \quad \text{and} \quad d = 0.03 \text{ m} = 3 \text{ cm}$$

**315.** $E = 5 \times 10^5$ V/m $= 5 \times 10^5$ N/C. For equilibrium, $F = Eq = mg$

$$5\times10^5 q = 3.2\times10^{-13} \quad q = 6.4\times10^{-19}\text{C} \quad \text{on the oil drop}$$

$$\frac{q}{e} = \frac{6.4\times10^{-19}}{1.6\times10^{-19}} = 4 \text{ electrons}$$

**316.** $E = V/d$. Set the upward electric force equal to the downward force of gravity.

$$F = Eq = mg \qquad \frac{Vq}{d} = mg$$

$$\frac{V(4\times1.60\times10^{-19})}{0.018} = (1.8\times10^{-15})(9.8) \qquad \text{and} \qquad V = 500 \text{ V}$$

**317.** **(A)** From $V = Ed$, we have $E = 12.0/3.00 \times 10^{-3} = 4000$ V/m.

**(B)** Since the battery was removed, the charge remains the same, as does $E$. Potential = $4000(5.00 \times 10^{-3}) = 20$ V.

**318.** We use the condition for equilibrium: sum of forces equals zero. Tension $T$ on the mass is up, weight $mg$ is down, and $qE$ can be up or down. We have $mg = (6.0 \times 10^{-4})(9.8) =$ 5.88 mN, $qE = (qV)/d = [(20\ \mu\text{C})(28\ \text{V})]/(0.10\ \text{m}) = 5.60$ mN so that $T = mg \pm qE =$ 11.5 MN or 0.28 MN.

**319.** Since the plate is 5.0 V lower, and $q = -e$ is negative, the electron is moving "uphill." The potential energy gained by the electron in reaching the plate is thus $eV = (1.6 \times 10^{-19}\,\text{C}) \times (5\ \text{V}) = 8.0 \times 10^{-19}$ J. Thus, there must be at least this much kinetic energy at the start, to supply the needed potential energy gain. $\text{KE} = \frac{1}{2}mv^2 \geq 8.0 \times 10^{-19}$ J, or

$$v \geq \left[\frac{16.0 \times 10^{-19}}{9.1 \times 10^{-31}}\right]^{1/2} \text{m/s} = 1330\ \text{km/s}$$

**320.** The energy $= eV = 20\ \text{keV} = (1.6 \times 10^{-19})(20000) = 3.2 \times 10^{-15}$ J. To get the speed, we have $3.2 \times 10^{-15}\,\text{J} = \frac{1}{2}(9.1 \times 10^{-31}\,\text{kg})v^2$. Solving, $v = 8.4 \times 10^7$ m/s. $v = 28$ percent of speed of light, so our nonrelativistic approximation is crude but still reasonable.

**321.** The negatively charged electron falls downhill toward the positive plate. The electric potential energy lost equals the gain in kinetic energy, so

$$\frac{1}{2}mv^2 = (1.6 \times 10^{-19}\,\text{C})(1.5\ \text{V}) = 2.4 \times 10^{-19}\,\text{J} \qquad v = \left[\frac{4.8 \times 10^{-19}}{9.1 \times 10^{-31}}\right]^{1/2} \text{m/s} = 730\ \text{km/s}$$

**322.** The proton will be repelled and caused to move radially toward infinity by the positive nucleus. In effect, it falls through the potential difference between point $P$ and infinity. Because the potential at infinity is defined as zero. This potential difference is just the absolute potential at point $P$;

$$V = \frac{1}{4\pi e_0}\frac{80 \times 1.6 \times 10^{-19}}{10^{-14}} = 12 \times 10^6\,\text{V}$$

In falling through this potential difference, the proton will thus acquire a kinetic energy of $K = 12$ MeV. In writing this, we assume the nucleus to remain nearly at rest. We then have

$$\frac{1}{2}mv^2 = (12 \times 10^6\,\text{eV})(1.602 \times 10^{-19}\,\text{J/eV})$$

Using the proton mass $1.67 \times 10^{-27}$ kg and solving for the proton's speed, we find that $v = 4.8 \times 10^7$ m/s.

**323.** The mass and charge of a proton are $1.67 \times 10^{-27}$ kg and $1.60 \times 10^{-19}$ C, respectively. When the proton is moved from plate $B$ to plate $A$, it loses a potential energy $q(V_B \times V_A)$, where $V_B - V_A$ is 100 V in this case. This appears as kinetic energy of the proton at plate $A$. The law of conservation of energy therefore tells us that

$$\text{loss in potential energy} = \text{gain in kinetic energy} \qquad \text{or} \qquad q(V_B - V_A) = \frac{1}{2}mv^2$$

Placing in the values and solving for $v$, we find 140 km/s.

**324. (A)** For a parallel-plate capacitor, $C=(e_0A)/d=[(8.85\times10^{-12})(0.02)]/(0.004)=$ 44 pF.

**(B)**

$$q=CV=(4.4\times10^{-11}\text{F})(500\text{ V})=22\text{ nC}$$

$$\text{energy}=\frac{1}{2}qV=\frac{1}{2}(2.2\times10^{-8}\text{C})(500\text{ V})=5.5\ \mu\text{J}$$

$$E=\frac{V}{d}=\frac{500\text{ V}}{4\times10^{-3}\text{m}}=125\text{ kV/m}$$

**(C)** The capacitor will now have a capacitance 2.60 times larger than before. Therefore, $q'=(2.60)(22)=57$ nC, and so $q'-q=35$ nC must flow onto it.

**325. (A)** $Q=CV=5\times10^{-6}(30)=150\ \mu\text{C}$.

**(B)** The charge on the plates remains the same when the oil replaces the air. The capacitance increases by a factor $K$.

$$C'=KC=10.5\ \mu\text{F}\qquad V'=\frac{V}{K}=14\text{ V}$$

**326. (A)** $$\frac{1}{C_{eq}}=\frac{1}{C_1}+\frac{1}{C_2}=\frac{1}{3\text{ pF}}+\frac{1}{6\text{ pF}}=\frac{1}{2\text{ pF}}$$

from which $C=2$ pF.

**(B)** In a series combination, each capacitor carries the same charge, which is the charge on the combination. Thus, using the result of (*A*),

$$q_1=q_2=q=C_{eq}V=(2\times10^{-12}\text{F})(1000\text{ V})=2\text{ nC}$$

**(C)** $$V_1=\frac{q_1}{C_1}=\frac{2\times10^{-9}\text{C}}{3\times10^{-12}\text{F}}=667\text{ V}\qquad V_2=\frac{q_2}{C_2}=\frac{2\times10^{-9}\text{C}}{6\times10^{-12}\text{F}}=333\text{ V}$$

**(D)** energy in $C_1=\frac{1}{2}q_1V_1=\frac{1}{2}(2\times10^{-9}\text{C})(667\text{ V})=6.7\times10^{-7}\text{ J}$
energy in $C_2=\frac{1}{2}q_2V_2=\frac{1}{2}(2\times10^{-9}\text{C})(333\text{ V})=3.3\times10^{-7}\text{ J}$
energy in combination $=(6.7+3.3)\times10^{-7}\text{ J}=10\times10^{-7}\text{ J}$

The last result is also directly given by $\frac{1}{2}qV$ or $\frac{1}{2}C_{eq}V^2$.

**327.** The equivalent capacitance of the three in series $=\frac{12}{13}=0.92\ \mu\text{F}$. Capacitors in series each carry the same charge, which is the same as the charge on the equivalent capacitor: $Q_{eq}=C_{eq}V=0.92(6)=5.5\ \mu\text{C}$. The $V$ across 4 $\mu$f is $V=Q_{eq}/C=5.5/4=1.4$ V.

**328. (A)** For the two capacitors in series,

$$\frac{1}{C_A}=\frac{1}{C_1}+\frac{1}{C_2}=\frac{1}{3}+\frac{1}{6}=\frac{2}{6}+\frac{1}{6}=\frac{3}{6}$$

and thus the capacitance of the upper branch is $C_A=2\ \mu\text{F}$. For the two parallel branches

$$C=C_A+C_B=2+4=6\ \mu\text{F}\qquad\text{(capacitance of the system)}$$

**(B)** Use the general formula $Q = CV$ successively for $C$ and for $C_B$.

$$Q = CV = 6\times10^{-6}(12) = 72\ \mu\text{C} \qquad \text{(charge on the system)}$$

$$Q_B = C_B V = (4\times10^{-6})(12) = 48\ \mu\text{C} \qquad \text{(charge on the 4-}\mu\text{F capacitor)}$$

The total charge on the system is the sum of the charges on the upper branch and on the lower branch; thus, $Q_A = Q - Q_B = 24\ \mu$C. Both the 3- and the 6-$\mu$F capacitors carry this same charge, since they are in series.

**(C)** The 4-$\mu$F capacitor has a potential difference of 12 V, the applied voltage. Use $Q = CV$ for the 3-$\mu$F capacitor.

$$Q_A = (3\times10^{-6})V_3 \qquad \text{or} \qquad 2.4\times10^{-5} = (3\times10^{-6})V_3 \qquad \text{so} \qquad V_3 = 8\ \text{V}$$

Use $Q = CV$ for the 6-$\mu$F capacitor.

$$Q_A = (6\times10^{-6})V_6 \qquad \text{or} \qquad 2.4\times10^{-5} = (6\times10^{-6})V_6 \qquad \text{and} \qquad V_6 = 4\ \text{V}$$

Their sum must, of course, equal the terminal voltage, and indeed 8 V + 4 V = 12 V.

**329. (A)** First find the capacitance $C_P$ of the parallel section.

$$C_P = C_2 + C_3 = 6 + 4 = 10\ \mu\text{F}$$

$C_P$ is in series with $C_1$, so

$$\frac{1}{C} = \frac{1}{C_1} + \frac{1}{C_P} = \frac{1}{5} + \frac{1}{10} = \frac{3}{10} \qquad \text{and} \qquad C = 3.3\ \mu\text{F}$$

**(B)** Find the total charge $Q$ on the system first.

$$Q = CV = 3.33\times10^{-6}(1000) = 3.3\times10^{-3}\text{C}$$

Then

$$V_1 = \frac{Q}{C_1} = \frac{3.3\times10^{-3}}{5\times10^{-6}} = 0.67\times10^3 = 670\ \text{V} \quad V_2 = V_3 = V - V_1 = 1000 - 670 = 330\ \text{V}$$

As a check, we can use the equivalent of the two parallel capacitors, $C_p$.

$$V_2 = V_3 = \frac{Q}{C_P} = \frac{3.33\times10^{-3}}{10\times10^{-6}} = 333\ \text{V}$$

**330.** Reduce the system as shown in Figure 330.1, to find $C_{eq} = 2.0\ \mu$F. The charge on the equivalent capacitor $Q = CV = 2(12) = 24\ \mu$C. This is also the charge on the equivalent 6-$\mu$F capacitor. So $V$ across it (and the 2 $\mu$F) is $Q/C = 24/6 = 4.0$ V.

**331.**

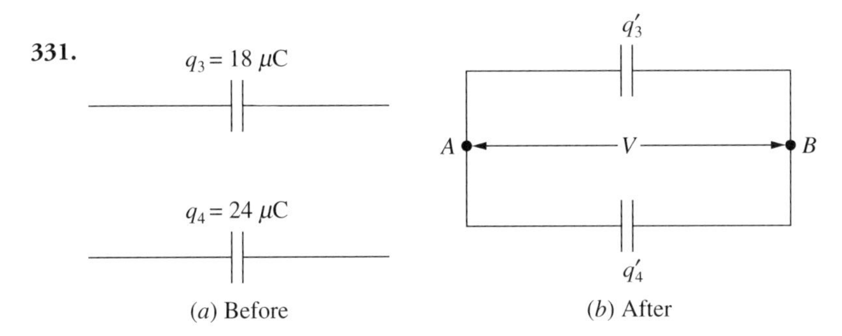


**Figure A20.2**

The situation is shown in Figure A20.2. Before being connected, their charges are

$$q_3 = CV = (3\times10^{-6}\,\text{F})(6\text{ V}) = 18\ \mu\text{C} \qquad q_4 = CV = (4\times10^{-6}\,\text{F})(6\text{ V}) = 24\ \mu\text{C}$$

As seen in the figure, the charges will partly cancel when the capacitors are connected together. Their final charges are given by $q_3' + q_4' = q_4 - q_3 = 6\mu\text{C}$. Also, the potential across each is now the same so that

$V = q/C$ gives

$$\frac{q_3'}{3\times10^{-6}\,\text{F}} = \frac{q_4'}{4\times10^{-6}\,\text{F}} \qquad \text{or} \qquad q_3' = 0.75q_4'$$

Substitution of this in the previous equation gives

$$0.75q_4' + q_4' = 6\ \mu\text{C} \quad \text{or} \quad q_4' = 3.43\ \mu\text{C}$$

Then $q_3' = 0.75q_4' = 2.57\ \mu\text{C}$.

**332.** Originally, $Q_1$ = 48 μC, $Q_2$ = 72 μC; hence, $Q_1' + Q_2' = 72 - 48 = 24\ \mu\text{C}$. Also, $V_1' = V_2'$ gives $Q_1'/C_1 = Q_2'/C_2$. Solving simultaneously gives 10 μC and 14 μC.

## Chapter 21: Simple Electric Circuits

**333.** First we determine the current, $I = Q/t$, or $I$ = 720 C/60 s = 12 A. Then use Ohm's law, $V = IR$, or $V$ = (12 A)(5 Ω) = 60 V.

**334.** $R = \rho(L/A)$. The resistivity can be looked up to be $\rho = 2.8\times10^{-8}\Omega\cdot\text{m}$. The cross-sectional area $A$ is $\pi r^2 = 3.14(0.5\times10^{-3}\text{m})^2 = 7.85\times10^{-7}\text{m}^2$. Then $4\Omega = [(2.8\times10^{-8}\Omega\cdot\text{m})L]/(7.85\times10^{-7}\text{ m}^2)$; and solving we get $L$ = 112 m.

**335.** The resistance $R$ of the parallel combination is given by

$$\frac{1}{R} = \frac{1}{R_1} + \frac{1}{R_2} + \frac{1}{R_3} = \frac{1}{12} + \frac{1}{16} + \frac{1}{20} = \frac{20}{240} + \frac{15}{240} + \frac{12}{240} = \frac{47}{240} \qquad \text{or} \qquad R = 5.11\ \Omega$$

Then

$$R_x + R = 25 \qquad \text{or} \qquad R_x = 25 - 5.11 = 19.9\ \Omega$$

**336.** Figure A21.1 shows all possible combinations and their equivalent resistances.

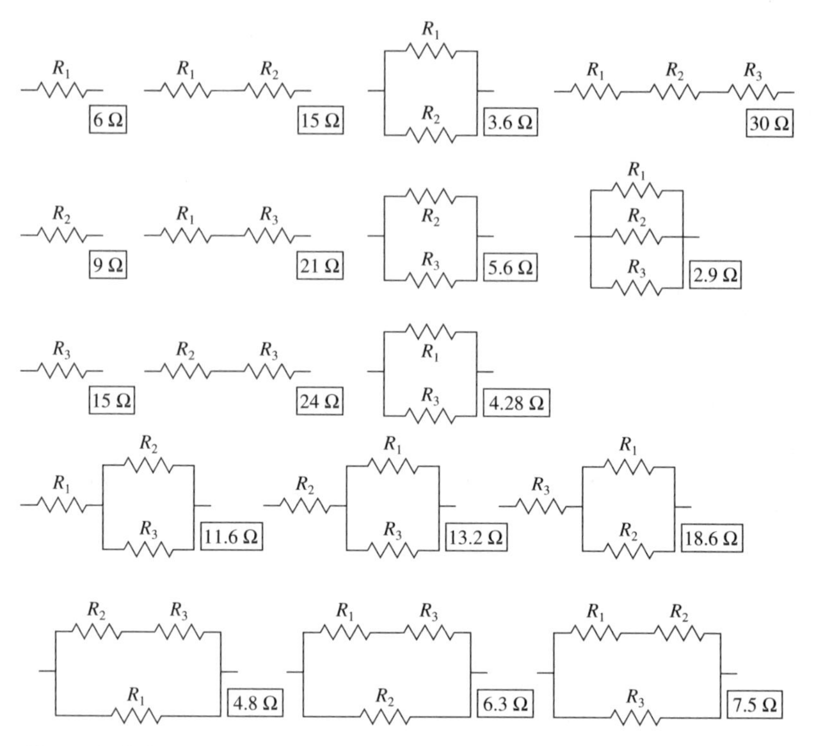


**Figure A21.1**

**337.** It is clear that we cannot have any resistance in series with the remaining pair. This observation leads to the solution indicated in Figure A21.2.

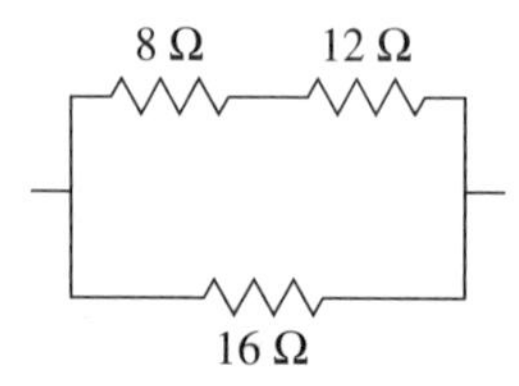


**Figure A21.2**

**338. (A)** The total resistance must be

$$R = \frac{E}{I} = \frac{45\ \text{V}}{0.45\ \text{A}} = 100\ \Omega$$

But $\frac{1}{R} = \frac{1}{R_1} + \frac{1}{R_2}$ so $\frac{1}{R_2} = \frac{1}{R} - \frac{1}{R_1} = \frac{1}{100\ \Omega} - \frac{1}{300\ \Omega} = \frac{2}{300\ \Omega}$ or $R_2 = 150\ \Omega$

**(B)** $I_1 = \frac{E}{R_1} = \frac{45\ \text{V}}{300\ \Omega} = 0.15\ \text{A}$ $\quad I_2 = \frac{E}{R_2} = \frac{45\ \text{V}}{150\ \Omega} = 0.30\ \text{A}$

**339.** **(A)** The sum of $R_2$ and $R_3$ in parallel is

$$\frac{1}{R'} = \frac{1}{R_2} + \frac{1}{R_3} = \frac{1}{50\ \Omega} + \frac{1}{100\ \Omega} = \frac{3}{100\ \Omega} \qquad \text{or} \qquad R' = 33.3\ \Omega$$

Since $R'$ is in series with $R_1$, the total resistance of the circuit is $R = R' + R_1 = 33.3\ \Omega + 25\ \Omega = 58\ \Omega$.

**(B)** $I = \dfrac{E}{R} = \dfrac{12\ \text{V}}{58.3\ \Omega} = 0.206\ \text{A}$

The potential $V'$ across $R_2$ and $R_3$ is $V' = E - R_1 \quad I = 12\ \text{V} - (25\ \Omega)(0.206\ \text{A}) = 6.85\ \text{V}$. Therefore,

$$I_2 = \frac{V'}{R_2} = \frac{6.85\ \text{V}}{50\ \Omega} = 0.137\ \text{A} \qquad I_3 = \frac{V'}{R_3} = \frac{6.85\ \text{V}}{100\ \Omega} = 0.0685\ \text{A}$$

**340.** **(A)** The sum of $R_2$ and $R_3$ in series is $R' = R_2 + R_3 = 25\ \Omega + 15\ \Omega = 40\ \Omega$. Since $R'$ is in parallel with $R_1$, the total resistance of the circuit is

$$\frac{1}{R} = \frac{1}{R_1} + \frac{1}{R'} = \frac{1}{80\ \Omega} + \frac{1}{40\ \Omega} = \frac{3}{80\ \Omega} \qquad \text{or} \qquad R = 26.7\ \Omega$$

**(B)** $E = R_1 I_1 = (80\ \Omega)(0.3\ A) = 24\ \text{V} \qquad I_2 = \dfrac{E}{R'} = \dfrac{24\ \text{V}}{40\ \Omega} = 0.6\ \text{A}$

$$I = I_1 + I_2 = 0.9\ \text{A}$$

**341.** **(A)** $R = R_1 + R_2 + R_3 = 45\ \Omega \qquad I_1 = I_2 = I_3 = I = \dfrac{\varepsilon}{R} = \dfrac{3.0\ \text{V}}{45\ \Omega} = 0.067\ \text{A}$

**(B)** $I_1 = \dfrac{\varepsilon}{R_1} = \dfrac{3.0\ \text{V}}{5.0\ \Omega} = 0.6\ \text{A} \qquad I_2 = \dfrac{\varepsilon}{R_2} = \dfrac{3.0\ \text{V}}{15.0\ \Omega} = 0.2\ \text{A}$

$I_3 = \dfrac{\varepsilon}{R_3} = \dfrac{3.0\ \text{V}}{25\ \Omega} = 0.12\ \text{A} \qquad I = I_1 + I_2 + I_3 = 0.92\ \text{A}$

**(C)** $I_3 = \dfrac{\varepsilon}{R_3} = \dfrac{3.0\ \text{V}}{25\ \Omega} = 0.12\ \text{A} \qquad I_1 = I_2 = \dfrac{\varepsilon}{R_1 + R_2} = \dfrac{3.0\ \text{V}}{20\ \Omega} = 0.15\ \text{A}$

$I = I_1 + I_3 = 0.27\ \text{A}$

**(D)** $\dfrac{1}{R'} = \dfrac{1}{R_2} + \dfrac{1}{R_3} = \dfrac{1}{15} + \dfrac{1}{25} = \dfrac{8}{75} \qquad R' = 9.4\ \Omega$

$I_1 = I = \dfrac{\varepsilon}{R_1 + R'} = \dfrac{3.0\ \text{V}}{14.375\ \Omega} = 0.209\ \text{A}$

$I_2 = \dfrac{V_2}{R_2} = \dfrac{\varepsilon - I_1 R_1}{R_2} = \dfrac{1.96\ \text{V}}{15\ \Omega} = 0.130\ \text{A} \qquad I_3 = I - I_2 = 0.079\ \text{A}$

**342.** Let $V$ be the potential difference $V_b - V_a$. Applying Ohm's law, $I_1 = \dfrac{\varepsilon}{R + r} = \dfrac{3.0\ \text{V}}{5.5\ \Omega} = 0.545\ \text{A}$. Then $V = \varepsilon - rI = (3.0\ \text{V}) - (0.5\ \Omega)(0.545\ \text{A}) = 2.7\ \text{V}$

**343.** **(A)** For the two cases,

$$I_1 = \frac{\%}{R_1 + r} \quad \text{and} \quad I_2 = \frac{\%}{R_2 + r} \quad \text{so} \quad \frac{I_1}{I_2} = \frac{R_2 + r}{R_1 + r} \quad \text{or} \quad \frac{20\ \Omega + r}{10\ \Omega + r} = \frac{0.5\ \text{A}}{0.27\ \text{A}} = 1.85$$

Solving this last equation for $r$, we get

$$20\ \Omega + r = 1.85(10\ \Omega + r) = 18.5\ \Omega + 1.85r \quad r = \frac{1.5\ \Omega}{0.85} = 1.8\ \Omega$$

**(B)** $\varepsilon = I(R + r) = (0.5\ \text{A} \times 11.76\ \Omega) = 5.9\ \text{V}$

**344.** We have $V = E - ir$, with $E = 120$ V, $V = 110$ V, $i = 20$ A. Then $110 = 120 - 20r$, and $r = 0.5\ \Omega$.

**345.** Since the battery is charging, $V = E + Ir$, where $I = 10$ A, $E = 5.6$ V, and $V = 6.8$ V. Then $6.8 = 5.6 + 10r$, and $r = 0.12\ \Omega$.

**346.** $V_t = \varepsilon - Ir = 6 - 200(0.01) = 4$ V

**347.** Power $= VI = I^2R = (0.50)^2(200) = 50$ W, and this lost power will appear as heat in the resistor. In this case, 50 J of energy is lost each second, and so 50/4.184, or about 12 cal of heat, is generated each second.

**348.** Because power $= VI$, we can find $I$ as follows:

$$I = \frac{\text{power}}{V} = \frac{90\ \text{W}}{120\ \text{V}} = 0.75\ \text{A}$$

Since the potential drop across the bulb is 120 V and since the current through it is 0.75 A Ohm's law tells us that

$$R = \frac{V}{I} = \frac{120\ \text{V}}{0.75\ \text{A}} = 160\ \Omega$$

**349.** $P_{4\Omega} = \dfrac{V^2}{R} = \dfrac{(12\ \text{V})^2}{4\ \Omega} = 36\ \text{W} \quad P_{2\Omega} = \dfrac{(12\ \text{V})^2}{2\ \Omega} = 72\ \text{W}$

Since the brightness of a bulb increases with the power, the 2-Ω bulb is brighter than the 4-Ω bulb.

**350.** The electric energy generated is $Pt = 500t$ J, where $t$ is time in seconds. To heat the water requires thermal energy: $mc\,\Delta T = (0.250\ \text{kg})(4.184\ \text{kJ/kg}\cdot\text{K})(80\ \text{K}) = 83.7\ \text{kJ}$. Thus, $0.500t_{\text{min}} = 83.7$, or $t_{\text{min}} = 167$ s.

**351.** First, find the energy input necessary to charge the water temperature using $Q = mc\,\Delta T$. The mass of the coffee is ~ 200 g, because coffee is essentially water, and has density $\frac{1\text{g}}{\text{mL}}$.

$$Q = (0.200\ \text{kg})\left(4200 \frac{\text{J}}{\text{kg}\cdot{}^\circ\text{C}}\right)(70^\circ\text{C})$$

$$Q = 59000\ \text{J}$$

Now, the power necessary is $\frac{\text{energy}}{\text{time}} = \frac{59{,}000\text{ J}}{30\text{S}} = 2000$ watts

and then electrical power = $IV$

so $$I = \frac{P}{V} = \frac{2000\text{ watts}}{120\text{ volts}} = 17\text{ Amps}$$

**352.** *Kirchoff's Junction Rule*: The sum of all the currents coming into a junction must equal the sum of all the currents leaving the point.

*Kirchhoff's Loop Rule*: As one traces out a closed circuit loop, the algebraic sum of the potential changes encountered is zero. In this sum, a potential rise is positive and a potential drop is negative.

**353. (A)** When $k$ is open, $I_3 = 0$, because no current can flow through the open switch. Apply the point rule to point $a$.

$$I_1 + I_3 = I_2 \qquad \text{or} \qquad I_2 = I_1 + 0 = I_1$$

Apply the loop rule to loop *acbda*. In volts,

$$-12 + 7I_1 + 8I_2 + 9 = 0 \qquad (1)$$

To understand the signs used, remember that current always flows from high to low potential through a resistor.

Because $I_2 = I_1$, (1) becomes

$$15I_1 = 3 \quad \text{or} \quad I_1 = 0.20\text{ A}$$

Also, $I_2 = I_1 = 0.20$ A.

**354.** We refer to Problem 353. With $k$ closed, $I_3$ is no longer known to be zero. Applying the point rule to point $a$ gives

$$I_1 + I_3 = I_2 \qquad (2)$$

Applying the loop rule to loop *acba* gives, in volts,

$$-12 + 7I_1 - 4I_3 = 0 \qquad (3)$$

and to loop *adba* gives

$$-9 - 8I_2 - 4I_3 = 0 \qquad (4)$$

Applying the loop rule to the remaining loop, *acbda*, would yield a redundant equation, because, among the three loop equations, each voltage change would appear twice.

## Chapter 22: The Magnetic Field

**355.** $F = qvB \sin \theta$; direction follows right-hand rule.

**(A)** **v** is perpendicular to **B**, so $\theta = 90°$ and $F = qvB$. Direction (rotating **v** arrowhead toward **B**) is into paper.

**(B)** Here $F = qvB \sin(\pi - \theta) = qvB \sin \theta$. Rotating **v** toward **B** through the smallest possible angle, we have **F** out of paper.

**(C)** Here **v** is perpendicular to **B**, so $F = qvB$. Rotating **v** into paper gives **F** in plane of paper 90° counterclockwise from **v**.

**356.** Right (direction of progress of right-hand screw rotating from **v** into **B**).

**357.** Since the magnetic force is always perpendicular to **B** and to **v**, it acts as a centripetal force. Furthermore, since there is no acceleration tangent to the path of motion after passing through the potential difference, both $v$ and $F = evB$ are constant, which are the conditions for uniform circular motion. First, we find $v$ from $eV = \frac{1}{2}mv^2$, or $(1.6 \times 10^{-19}\text{ C})(3750\text{ V}) = \frac{1}{2}(9.1 \times 10^{-31}\text{ kg})v^2$, or $v = 3.63 \times 10^7$ m/s. Then from Newton's second law, $qvB = (mv^2)/R$, where $R$ is the radius of the circular motion. Then $R = (mv)/(qB) = [(9.1 \times 10^{-31}) \times (3.63 \times 10^7)]/[(1.6 \times 10^{-19})(4 \times 10^{-3}\text{ T})] = 52$ mm.

**358.** The KE of a particle is conserved in the magnetic field:

$$\frac{1}{2}mv^2 = Vq \qquad \text{or} \qquad v = \sqrt{\frac{2Vq}{m}}$$

They follow a circular path in which

$$r = \frac{mv}{qB} = \frac{m}{qB}\sqrt{\frac{2Vq}{m}} = \frac{1}{B}\sqrt{\frac{2Vm}{q}} = \frac{1}{0.2\text{ T}}\sqrt{\frac{2(1000\text{ V})(6.68 \times 10^{-27}\text{ kg})}{3.2 \times 10^{-19}\text{C}}} = 32\text{ mm}$$

**359.** To just miss the opposite plate, the particle must move in a circular path with radius $d$, so from $Bqd = mv$, and using $K = (mv^2)/2$, we have $B = (2mK)^{1/2}/(qd)$.

**360.** We resolve the particle velocity into components parallel to and perpendicular to the magnetic field. The magnetic force due to $v_{\parallel}$ is zero ($\sin\theta = 0$); the magnetic force due to $v_{\perp}$ has no $x$ component. Therefore, the $x$ motion is uniform, at speed $v_{\parallel} = (0.86) \times (8 \times 10^6\text{ m/s}) = 6.88 \times 10^6$ m/s, while the transverse motion in circular, with radius

$$r = \frac{mv_{\perp}}{qB} = \frac{(1.67 \times 10^{-27}\text{ kg})(0.5 \times 8 \times 10^6\text{ m/s})}{(1.6 \times 10^{-19}\text{C})(0.15\text{ T})} = 0.28\text{ m}$$

The proton will spiral along the $x$ axis; the radius of the spiral (or helix) will be 28 cm.

To find the *pitch* of the helix (the $x$ distance traveled during one revolution), we note that the time taken to complete one circle is

$$\text{period} = \frac{2\pi r}{v_{\perp}} = \frac{2\pi(0.28\text{ m})}{(0.5)(8 \times 10^6\text{ m/s})} = 4.4 \times 10^{-7}\text{ s}$$

During that time, the proton will travel an $x$ distance of

$$\text{pitch} = (v_{\parallel})(\text{period}) = (6.88 \times 10^6\text{ m/s})(4.4 \times 10^{-7}\text{ s}) = 3.0\text{ m}$$

**361.** The vertical component of **B** is parallel to the current and does not contribute to the force; therefore, $F = ILB_{\text{H}} = (30\text{ A})(1\text{ m})(2 \times 10^{-5}\text{ T}) = 6 \times 10^{-4}$ N, west.

**362.** From the RHR, the magnetic force $ILB$ is directed downward. This constant force shifts the equilibrium position downward by $(ILB)/(2k)$. (Two springs in parallel, each of constant $k$, are equivalent to one spring of constant $2k$.)

**363.** **(A)** $B_1 = (2\times10^{-7}\,\text{H/m})\dfrac{8\text{ A}}{0.030\text{ m}} = 53.3\,\mu\text{T}$

$$B_2 = (2\times10^{-7}\,\text{H/m})\frac{12\text{ A}}{0.150\text{ m}} = 16.0\,\mu\text{T}$$

$\mathbf{B}_1$ and $\mathbf{B}_2$ are in opposite directions at $A$, so the magnitude of their sum is $B = B_1 - B_2 = 37.3\,\mu$T.

**(B)** The total field is zero at the point between the wires where

$$\frac{I_1}{r_1} = \frac{I_2}{r_2} \quad \text{or} \quad \frac{r_2}{r_1} = \frac{I_2}{I_1} = \frac{12\text{ A}}{8\text{ A}} = 1.5$$

Since $r_1 + r_2 = 180$ mm, we obtain $r_1 = 72$ mm.

**364.** The westward deflection of the compass needle indicates that the field *above the wire* must be directed toward the west. By the right-hand rule, in order to produce this field, the current in the wire must flow from north to south.

**365.**

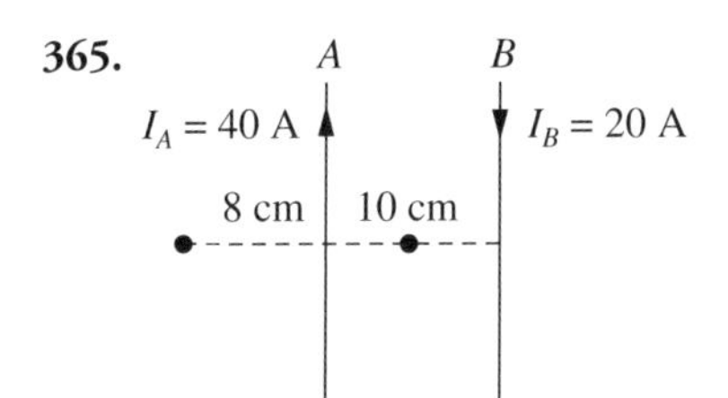


**Figure A22.1**

**(A)** At the midpoint between the wires (Figure A22.1) the fields both point into the page and hence reinforce:

$$B = B_A + B_B = \frac{(2\times10^{-7})(40+20)}{0.05} = 2.4\times10^{-4}\,\text{T}$$

**(B)** At the left-hand dot the field due to A points out of the page, and the field due to B points into the page. These fields will thus be subtracted.

$$B = B_A - B_B = \frac{(2\times10^{-7})(40)}{(0.08)} - \frac{(2\times10^{-7})(20)}{(0.18)} = 7.8\times10^{-5}\,\text{T}$$

**366.** The fields due to wires $D$ and $G$ at wire $C$ are

$$B_D = \frac{\mu_0 I}{2\pi r} = \frac{(4\pi\times10^{-7}\,\text{T}\cdot\text{m/A})(30\text{ A})}{2\pi(0.03\text{ m})} = 2\times10^{-4}\,\text{T}$$

into the page, and

$$B_G = \frac{(4\pi\times10^{-7}\,\text{T}\cdot\text{m/A})(20\text{ A})}{2\pi(0.05\text{ m})} = 0.8\times10^{-4}\,\text{T}$$

out of the page. Therefore, the field at the position of wire $C$ is

$$B = 2\times10^{-4} - 0.8\times10^{-4} = 1.2\times10^{-4}\,\text{T}$$

into the page. The force on a 25-cm length of $C$ is

$$F = ILB\sin\theta = (10\text{ A})(0.25\text{ m})(1.2\times10^{-4}\,\text{T})(\sin 90°) = 3\times10^{-4}\,\text{N}$$

Using the right-hand rule at wire $C$ tells us that the force on wire $C$ is toward the right.

**367.** Field at $MN$ due to $I_1$ is $(\mu_0 I_1)/(2\pi a)$. The force on $MN$ due to $I_1$ is away from $I_1$ and is $I_2LB = (\mu_0 I_1 I_2 L)/(2\pi a)$. Forces on the loop due to $I_1$ act in the plane of the loop, giving zero torque. These forces compress the loop.

## Chapter 23: Induced emf; Generators and Motors

**368.** Let $\hat{n}$ represent the outward unit normal vector to a given face of the box and $A$ the area of that face. Then the outward flux through the face is $\Phi = \mathbf{B} \cdot \hat{\mathbf{n}}A = BA\cos\theta$, where $\theta =$ angle $(\mathbf{B}, \hat{\mathbf{n}})$.

Clearly, $\Phi = 0$ through the two side faces ($\hat{\mathbf{n}}$ in $\pm z$ direction) and the bottom face ($\hat{\mathbf{n}}$ in $-y$ direction). Through the front and back faces, $\hat{\mathbf{n}}$ is along $+x$ and $-x$, respectively, so

$$\Phi_{\text{front}} = (0.2\ \text{T})(40 \times 10^{-4}\ \text{m}^2)(1) = 0.8\ \text{mWb}$$
$$\Phi_{\text{back}} = (0.2\ \text{T})(90 \times 10^{-4}\ \text{m}^2)(-1) = -1.8\ \text{mWb}.$$

(The minus sign indicates flux is inward through surface.) For the top surface $\theta = 60°$ and $\Phi_{\text{top}} = (0.2\ \text{T})(100 \times 10^{-4}\ \text{m}^2)(1/2) = 1.0\ \text{mWb}$.

**369.** We know that $\Phi = B_{\perp}A$.

**(A)** $\Phi = B_{\perp}A = BA = (0.16\ \text{T})(15 \times 10^{-4}\ \text{m}^2) = 240\ \mu\text{Wb}$

**(B)** $\Phi = (B\cos 20°)A = (2.4 \times 10^{-4}\ \text{Wb})(\cos 20°) = 226\ \mu\text{Wb}$

**(C)** $\Phi = (B\sin 20°)A = (2.4 \times 10^{-4}\ \text{Wb})(\sin 20°) = 82\ \mu\text{Wb}$

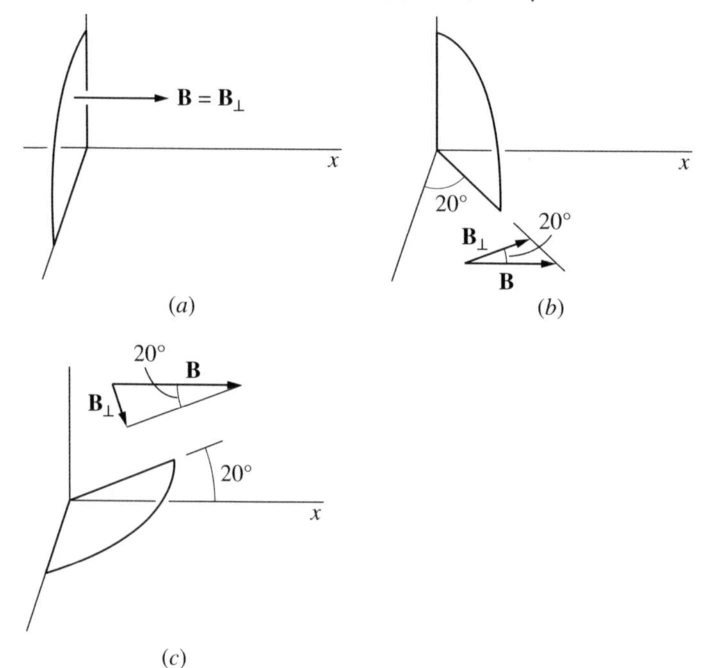


**Figure A23.1**

**370.** The initial flux $\phi_1$ is zero, and the final flux $\phi_2$ is

$$\phi_2 = NB_n A = (10)(18\ \text{T})(25\pi \times 10^{-4}\,\text{m}^2) = 1.41\ \text{Wb}$$

and so $\Delta\phi = \phi_2 - \phi_1 = 1.41$ Wb. The induced emf is

$$\mathscr{E} = -\frac{\Delta\phi}{\Delta t} = -\frac{1.41\ \text{Wb}}{3\ \text{s}} = -0.47\ \text{V}$$

The minus sign indicates that the emf will cause a current that creates a field opposing the change in **B** (see Figure 23.3). The magnitude of the induced current is

$$I_i = \frac{\mathscr{E}}{R} = \frac{0.47\ \text{V}}{2\ \Omega} = 0.235\ \text{A}$$

**371.** $\Delta\Phi = B_{\text{final}}A - B_{\text{initial}}A = (0.25\ \text{T})(\pi r^2) = (0.25\ \text{T})\pi(0.030\ \text{m})^2 = 7.1\times 10^{-4}\,\text{Wb}$ for each loop of the coil

$$|\mathscr{E}| = N\left|\frac{\Delta\Phi}{\Delta t}\right| = (50)\frac{7.1\times 10^{-4}\,\text{Wb}}{2\times 10^{-3}\,\text{s}} = 18\ \text{V}$$

**372.** The magnetic field inside the loop in Figure 372.1 is directed into the paper. As the loop is pulled away from the wire to regions of weaker field, the magnetic flux through the loop decreases in magnitude. As a consequence of Lenz's law, the magnetic field produced by the induced current must counteract the decrease in flux. Therefore, it must be directed into the plane of the figure (within the loop). Hence, the induced current must be *clockwise*.

**373.** The closing of the circuit is accompanied by a rapid increase in the magnitude of the magnetic flux through the ring, which induces a current in it. According to Lenz's law, the magnetic field due to the induced current must be directed opposite to the rapidly strengthening field due to the coil. This implies that the current in the ring circulates in the sense *opposite* to the current in the windings of the coil. But two opposite dipoles are like two antiparallel bar magnets: They repel each other, sending the ring upward.

**374.** A current is induced in the ring as it enters the magnetic field. By Lenz's law, the magnetic force on the induced current opposes the motion that induced the current. It is this force that halts the translational motion of the ring. The kinetic energy of the macroscopic motion is converted to thermal energy by Joule heating in the ring. Note that (other things being equal) the more highly conducting the ring material is, the more rapidly the motion is stopped and the energy is dissipated.

**375.** If the ring is cut, no current can flow in it. Hence, there is no magnetic force to halt the translation and no Joule heating to dissipate the ring's kinetic energy. The pendulum will swing freely until ordinary friction dissipates the energy.

**376. (A)** Consider first the coil on the left. As the magnet moves to the right, the flux through the coil, which is directed generally to the left, decreases. To compensate for this, the induced current in the coil will flow so as to produce a flux toward the left through itself.

Apply the right-hand rule to the loop on the left end. For it to produce flux inside the coil directed toward the left, the current must flow through the resistor from $B$ to $A$.

Now consider the coil on the right. As the magnet moves toward the right, the flux inside the coil, also generally to the left, increases. The induced current in the coil will produce a flux toward the right to cancel this increased flux. Applying the right-hand rule to the loop on the right end, we find that the loop generates flux to the right inside itself if the current flows from $C$ to $D$ through the resistor.

**(B)** In this case, the flux change caused by the magnet's motion is opposite to what it was in (**A**). Using the same type of reasoning, we find that the induced currents flow through the resistors from $A$ to $B$ and from $D$ to $C$.

**377.** The situation is similar to that in Problem 376, except that the coil on the right is oppositely wound. Here, both coils experience a decrease in flux to the left as the magnet rotates from the parallel position shown to the perpendicular position. The induced currents in both coils thus tend to create a field toward the left so the currents flow from $B$ to $A$ and from $C$ to $D$.

**378. (A)** $\Phi = BA\cos(\mathbf{B}, \mathbf{A}) = (0.8\ \text{Wb/m}^2)(75\times10^{-3})^2\ \text{m}^2] = 4.5\ \text{mWb}$

**(B)** $|\varepsilon_{\text{avg}}| = \dfrac{\Delta\Phi}{\Delta t} = \dfrac{4.5\times10^{-3}\ \text{Wb}}{1.5\times10^{-2}\ \text{s}} = 0.3\ \text{V}$

**379.** Use Faraday's law $\mathscr{E} = -[(N\ \Delta\Phi)/\Delta t]$.

$$\mathscr{E} = -\frac{(150\times10^{-4})(100)(0.001-0.0)}{0.1} = -0.015\ \text{V}$$

Note that in this example the induced emf is constant within the time interval considered.

**380.** From Faraday's law, the magnitude of average induced emf is

$$\mathscr{E}_{\text{avg}} = \frac{(275)(0.024)(8\times10^{-5})}{0.025} = 0.0211\ \text{V}$$

$$\Phi = BA = (8\times10^{-5}\ \text{Wb/m}^2)(0.024\ \text{m}^2) = 1.9\ \mu\text{Wb}$$

**381.** A wire of length $l$ moving at a velocity $v$ sweeps out an area $lv$ per second. The flux cut per second is equal to $BA/t$, or $Blv$; thus, $\mathscr{E} = (0.003\ \text{Wb/m}^2)(0.19\ \text{m})(11.5\ \text{m/s}) = 6.6\ \text{mV}$. For the case shown, Lenz's law implies the emf is counterclockwise through the imaginary circuit; hence, it points from bottom to top in the conductor.

*Another method* Every charge $q$ in the wire is moving with velocity $\mathbf{v}$ and experiences a force $F = qvB$, which by the right-hand rule points upward in the conductor for positive $q$. The work per unit charge (voltage), if these charges were free to move a distance $l$ along the conductor, is $V = (Fl)/q = Blv$.

**382.** The induced current is $BLv/R = (0.15\ \text{T})(0.50\ \text{m})(2\ \text{m/s})/(3\Omega) = 0.050\ \text{A}$. Then the magnetic force on the rod is $ILB = (0.050\ \text{A})(0.50\ \text{m})(0.15\ \text{T}) = 0.0038\ \text{N}$ or 3.8 mN.

**383.** When the generator is not connected, there is no current in its coils, and the effort required to turn it is only that required to overcome mechanical friction in its bearings. However, when the generator delivers current, there are magnetic forces on the wires in the coil due to the presence of the induced current in the magnetic field. By Lenz's law, the direction of the induced current is such that the magnetic torque on the coils opposes the rotation of the coils. The torque is proportional to the current and is therefore large if the load has low resistance.

**384.** Basically, $|\mathscr{E}| \propto \Delta\Phi/\Delta t$. If the flux per pole is doubled, all else being equal, $\Delta\Phi$ doubles, so $|\mathscr{E}|$ doubles. If the speed of the armature doubles, $\Delta t$ halves, and all else being equal, $|\mathscr{E}|$ doubles.

**385. (A)** Since the slide wire is perpendicular to the magnetic field, the magnetic force on the wire has magnitude $F_m = iBl$. The magnetic force is balanced by a leftward force of 0.25 N, so we have $i = F_m/(Bl) = 0.25/[(0.50)(0.25)] = 2.0$ A.

**(B)** With a current $i = 2.0$ A, the voltage drop $V_R$ across the resistor is $V_R = iR = (2.0)(1.0) = 2.0$ V.

**(C)** The back emf $V_b = V_0 - V_R = 6.0 - 2.0 = 4.0$ V.

**386. (A)** The induced current is $BLv/R$. The magnetic force on the rod is $ILB$. Substituting, the force of 0.25 N $= IL(BLv/R)$. Solving for $v$, $v = (0.25\text{ N})(1.0\ \Omega)/(6.0\text{ A}) \times (0.25\text{ m})^2(0.5\text{ T}) = 1.3$ m/s.

**(B)** The mechanical power is equal to force times speed, or (0.25 N)(1.3 m/s) = 0.33 W.

**(C)** The electrical power used is $IV = (6.0\text{ A})(6.0\text{ V}) = 36$ W. The efficiency is the power output/power input = (0.33 W)/(36 W) = 0.9%.

## Chapter 24: Electric Circuits

**387.** Consider the circuit shown in Figure A24.1. The capacitor is initially uncharged. If the switch is now closed, the current $i$ in the circuit and the charge $q$ on the capacitor vary, as shown in Figure A24.1. Writing the loop rule for this circuit gives, calling the voltage across the capacitor $V_c$, $-iR - V_c + \mathscr{E} = 0$, or $i = (\mathscr{E} - V_c)/R$.

At the first instant after the switch is closed, $V_c = 0$ and $i = \mathscr{E}/R$. As time goes on, $V_c$ increases and $i$ decreases. The time taken for the current to drop to $e^{-1} \approx 0.368$ of its initial value is $RC$, which is called the *time constant* of the $R$-$C$ circuit.

Also shown in Figure A24.1 is the variation of $q$, the charge on the capacitor, with time. At $t = RC$, $q$ has attained 0.632 of its final value.

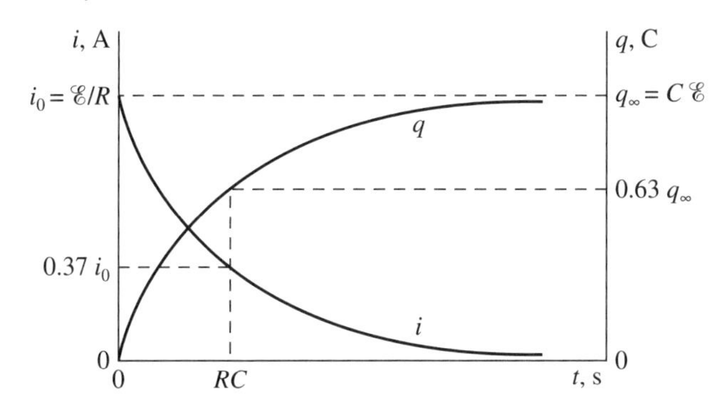


**Figure A24.1**

When a charged capacitor $C$ with initial charge $q_0$ is discharged through a resistor $R$, its discharge current follows the same curve as for charging. The charge $q$ on the capacitor follows a curve similar to that for the discharge current. At time $RC$, $i = 0.368i_0$ and $q = 0.368q_0$ during discharge.

**388. (A)** The loop rule applied to the circuit of Figure 387.1 at any instant gives $12\text{ V} - iR - V_c = 0$, where $V_c$ is the voltage across the capacitor. At the first instant, $q$ is essentially zero and so $V_c = 0$. Then $12\text{ V} - iR - 0 = 0$, or $i = 12\text{ V}/10^6\ \Omega = 12\ \mu\text{A}$.

**(B)** From Question 387, the current drops to 0.37 of its initial value when $t = RC = (10^6\ \Omega)(2 \times 10^{-6}\text{ F}) = 2\text{ s}$.

**(C)** At $t = 2$ s, the charge on the capacitor has increased to 0.63 of its final value. [See (*D*) below.]

**(D)** The final value for the charge occurs when $i = 0$ and $V_c = 12$ V. Therefore, $q_{\text{final}} = CV_c = (2 \times 10^{-6}\text{ F})(12\text{ V}) = 24\ \mu\text{C}$.

**389.** The loop equation for the discharging capacitor is $V_c - iR = 0$, where $V_c$ is the voltage across the capacitor. At the first instant, $V_c = 20$ kV, so

$$i = \frac{V_c}{R} = \frac{20\times10^3\text{ V}}{7\times10^6\,\Omega} = 2.9\text{ mA}$$

The voltage across the capacitor, as well as the charge on it, will decrease to 0.37 of its original value in one time constant. The required time is $RC = (7\times10^6\,\Omega)(5\times10^{-6}\text{ F}) = 35\text{ s}$.

**390.** From Question 387, $q$ decays to $0.37q_0$ in about one time constant. Since $V_c = q/C$, it drops in proportion to $q$. Then, from data given, $RC = 7$ s, or $C = (7\text{ s})/(10^4\ \Omega) = 700\ \mu\text{F}$.

**391.** When the switch in the circuit is first closed, the current in the circuit rises as shown in Figure A24.2. The current does not jump to its final value because the changing flux through the coil induces a back emf in the coil, which opposes the rising current. After $L/R$ seconds, the current has risen to 0.632 of its final value $i_\infty$. This time, $t = L/R$ is called the *time constant* of the *R-L* circuit. After a long time, the current is changing so slowly that the back emf in the inductor, $L(\Delta i/\Delta t)$, is negligible. Then $i = i_\infty = \mathscr{E}/R$.

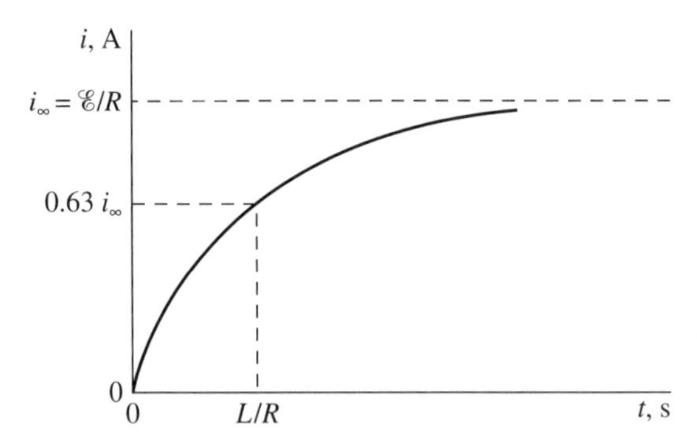


**Figure A24.2**

**392.** The time required is the time constant of the circuit.

$$\text{time constant} = \frac{L}{R} = \frac{1.5\ \text{H}}{0.6\ \Omega} = 2.5\ \text{s}$$

At long times, the current will be steady, and so no back emf will exist in the coil. Under such conditions,

$$I = \frac{\mathcal{E}}{R} = \frac{12\ \text{V}}{0.6\ \Omega} = 20\ \text{A}$$

**393.** The circuit is as shown in Figure A24.3 and obeys the loop equation $\mathcal{E} - L(\Delta i/\Delta t) - Ri = 0$. At $t = 0$, $i = 0$,

$$60 - 0.008\frac{\Delta i}{\Delta t} = 0 \qquad \frac{\Delta i}{\Delta t} = \frac{60}{0.008} = 7500\ \text{A/s}$$

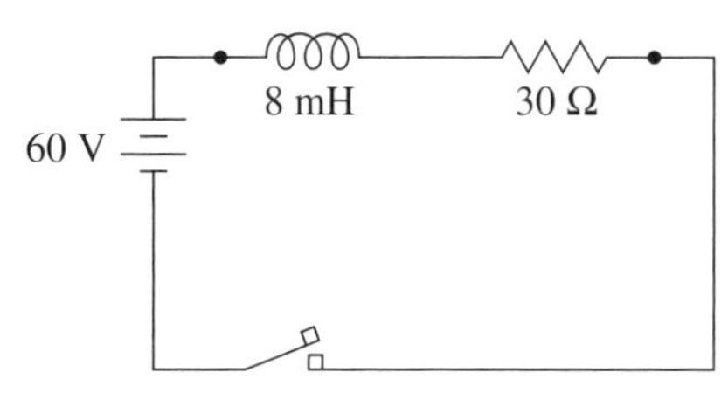


**Figure A24.3**

## Chapter 25: Light and Optical Phenomena

**394.** Since $n = c/v$, $v = c/n = 3\times10^8/2.4 = 1.25\times10^8$ m/s. As for the wavelength, $n = c/v = (\lambda_0 v_0)/(\lambda v) = \lambda_0/\lambda$, inasmuch as $v_0 = v$. Hence, $\lambda = 500/2.4 = 208$ nm.

**395.** **(A)** $v = \dfrac{c}{n} = \dfrac{3.0\times10^8\ \text{m/s}}{1.553} = 1.93\times10^8\ \text{m/s}$

**(B)** $n = \dfrac{c}{v} = \dfrac{3.0\times10^8\ \text{m/s}}{1.52\times10^8\ \text{m/s}} = 1.97$

**396.** **(A)** $\lambda' = \dfrac{v}{f} = \dfrac{c}{nf} = \dfrac{\lambda}{n}$

**(B)** $\lambda' = \dfrac{\lambda}{n} = \dfrac{420\ \text{nm}}{1.33} = 316\ \text{nm}$

**397.** Since $v = c/n$, $$t_1 = \frac{20}{c/n} = \frac{20(1.47)}{c} \quad \text{and} \quad t_2 = \frac{20(1.63)}{c}$$

Accordingly, $$t_2 - t_1 = \frac{20}{c}(1.63 - 1.47) = 1.07 \times 10^{-8}\,\text{s}$$

**398.** We use Snell's law: $n_1 \sin\theta_1 = n_2 \sin\theta_2$ with angles measured from normal. Then

$$\sin\theta_1 = \frac{n_2 \sin\theta_2}{n_1} = \frac{(1.36)(\sin 25°)}{1.00} = 0.574 \quad \text{or} \quad \theta_1 = 35°$$

**399.** $v = \dfrac{c}{n} = \dfrac{3\times 10^8}{1.33} = 2.25\times 10^8$ m/s

From Snell's law,

$$1 \sin 48° = 1.33 \sin \theta_w \quad \text{and} \quad \sin \theta_w = 0.5575;\ \theta_w = 34°$$

**400.** $v = \dfrac{c}{n} = \dfrac{3\times 10^8}{1.39} = 2.16\times 10^8$ m/s $\quad \dfrac{\sin 55°}{\sin\theta_{\text{al}}} = 1.39 \quad \sin\theta_{\text{al}} = 0.589 \quad \theta_{\text{al}} = 36°$

**401.**

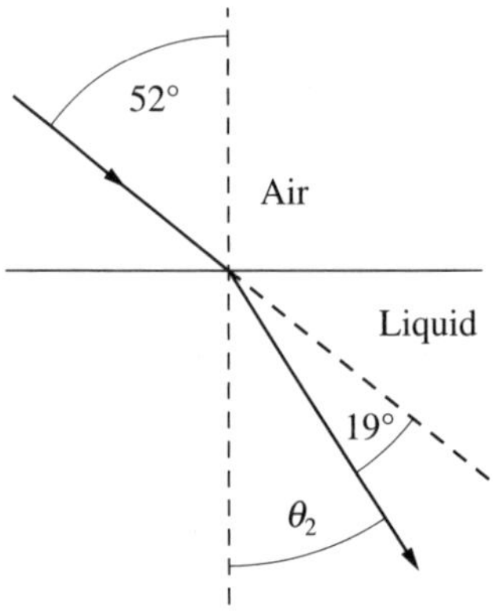

**Figure A25.1**

The situation is as shown in Figure A25.1.

$$\theta_2 = 52° - 19° = 33° \qquad n = \frac{\sin 52°}{\sin 33°} = \frac{0.7880}{0.5446} = 1.45$$

**402.** For the critical angle, $\sin\theta_c = n_2/n_1$, and so

**(A)** $\theta_c = \arcsin\dfrac{1}{1.76} = 34.6°$

**(B)** $\theta_c = \arcsin\dfrac{1.33}{1.76} = 49.1°$

**403.**

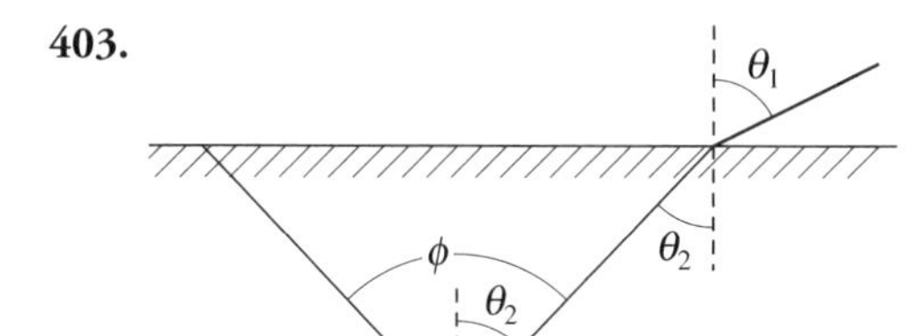


**Figure A25.2**

As can be seen from Figure A25.2, the maximum value of $\theta_2$ for observing light from the air corresponds to $\theta_1 = 90°$. Thus, $\theta_2$ is the critical angle and is given by

$$\sin\theta_2 = \frac{n_1 \sin 90°}{n_2} = \frac{(1.00)(1.0)}{1.333} = 0.75 \qquad \text{or} \qquad \theta_2 = 48.6°$$

The figure shows that $\phi = 2\theta_2$, so $\phi = (2)(48.5°) = 97.2°$.

**404.**

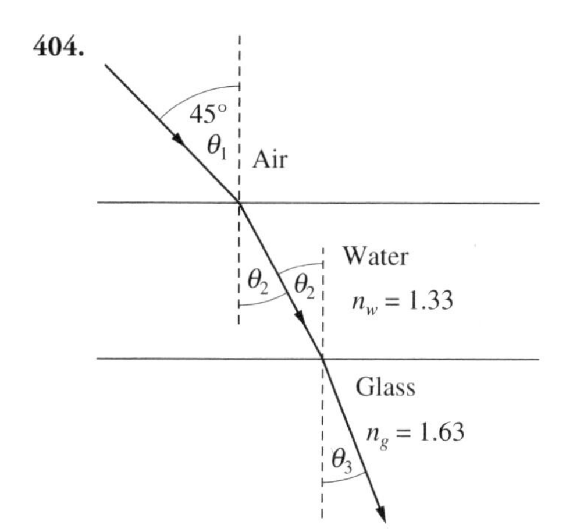


**Figure A25.3**

Refer to Figure A25.3. Use Snell's law for the air-water surface.

$$n_w = \frac{\sin\theta_1}{\sin\theta_2} \quad 1.33 = \frac{\sin 45°}{\sin\theta_2} = \frac{0.7071}{\sin\theta_2} \quad \sin\theta_2 = \frac{0.7071}{1.33} = 0.53 \quad \text{and} \quad \theta_2 = 32°$$

Now use Snell's law for the water-glass surface.

$$n_{BA} = \frac{n_g}{n_w} = \frac{\sin\theta_2}{\sin\theta_3} \quad \frac{1.63}{1.33} = \frac{0.53}{\sin\theta_3} \quad \sin\theta_3 = \frac{0.53(1.33)}{1.63} = 0.43 \quad \text{and} \quad \theta_3 = 26°$$

**405.**

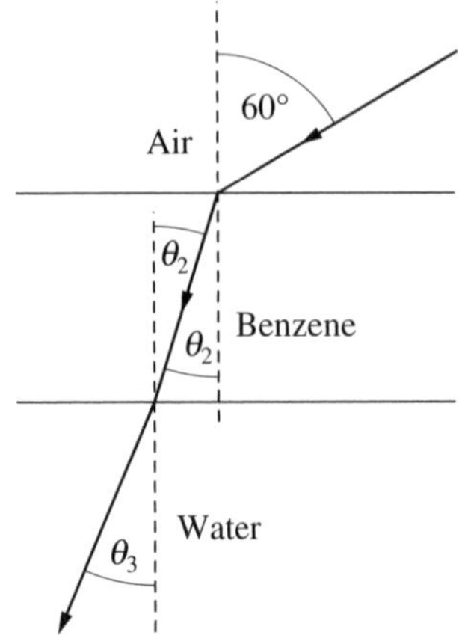


**Figure A25.4**

Here the light rays bend inward in the benzene and outward in the water, as shown in Figure A25.4. Use Snell's law first at the air-benzene surface and then at the benzene-water surface.

$$n_b = \frac{\sin\theta_1}{\sin\theta_2} \qquad 1.50 = \frac{\sin 60°}{\sin\theta_2} \qquad \sin\theta_2 = \frac{0.8660}{1.50} = 0.5773 \qquad \text{and} \qquad \theta_2 = 35°$$

Next

$$\frac{n_w}{n_b} = \frac{\sin\theta_2}{\sin\theta_3} \qquad \frac{1.33}{1.50} = \frac{0.577}{\sin\theta_3} \qquad \sin\theta_3 = \frac{0.577(1.50)}{1.33} = 0.65 \qquad \text{and} \qquad \theta_3 = 41°$$

**406. (A)** At the air-oil interface, sin 40° = 1.45 sin $\theta_{\text{oil}}$; at the oil-water interface, 1.45 sin $\theta_{\text{oil}}$ = 1.33 sin $\theta_w$, so $\theta_w = \sin^{-1}[(\sin 40°)/1.33] = 28.9°$.

**(B)** This is not possible, since $\theta_a > \theta_w$ and $\theta_a$ will reach 90° before $\theta_w$ does.

**(C)** In this case, sin 90° = 1.45 sin $\theta_{\text{oil}}$ = 1.33 sin $\theta_w$ so $(\theta_w)_{\text{critical}} = \sin^{-1}(1/1.33)$ = 48.6°, as for an air-water interface.

**407.**

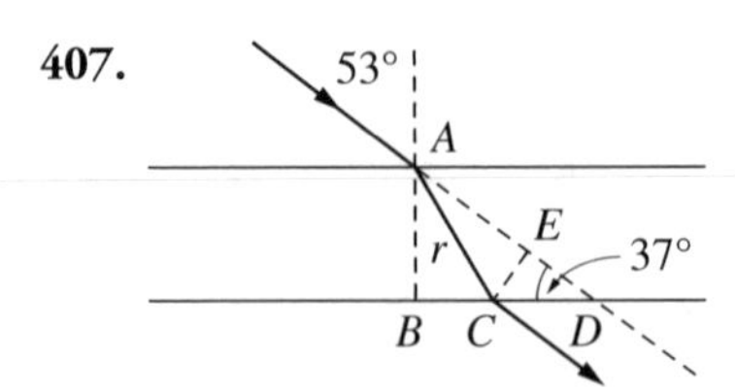


**Figure A25.5**

The situation is described by Figure. A25.5, with $CE$ the lateral displacement; $\overline{AB} = 20$ mm.

$$\sin r = \frac{\sin 53^\circ}{1.60} \quad r = 30^\circ \qquad \overline{BD} = 20 \tan 53^\circ = 26.5 \qquad \overline{BC} = 20 \tan 30^\circ = 11.5$$

$$\overline{CD} = 26.5 - 11.5 = 15.0 \qquad \overline{CE} = \overline{CD} \cos 53^\circ = 9.0 \text{ mm}$$

**408.**

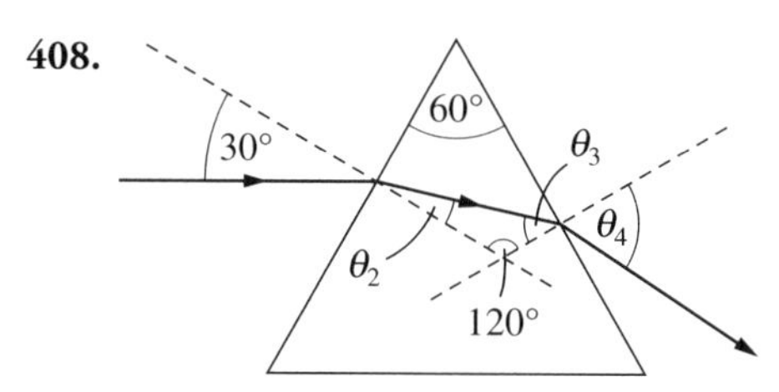


**Figure A25.6**

Analyze the refraction in sequence, from the left in Figure A25.6.

$$n = \frac{\sin\theta_1}{\sin\theta_2} \qquad 1.50 = \frac{\sin 30^\circ}{\sin\theta_2} \qquad \sin\theta_2 = \frac{0.5}{1.50} = 0.333 \quad \text{and} \quad \theta_2 = 20^\circ$$

As can be seen from the figure, $\theta_2 + \theta_3 + 120^\circ = 180^\circ$, or $\theta_2 + \theta_3 = 60^\circ$, the prism angle. Then $20^\circ + \theta_3 = 60^\circ$, and $\theta_3 = 40^\circ$. Next

$$n = \frac{\sin\theta_4}{\sin\theta_3} \qquad 1.50 = \frac{\sin\theta_4}{\sin 40^\circ} \quad 0.649(1.50) = \sin\theta_4 \quad \text{and} \quad \theta_4 = 77^\circ$$

**409.** Here we cannot use the small-angle approximation, or the minimum deviation formula. Instead, we go back to basics and follow the procedure of Question 408. For the case with $n_2 = 1.65$, Snell's law applied to the first surface gives

$$\sin\theta_2 = \frac{n_1 \sin\theta_1}{n_2} = \frac{(1.00)(\sin 65^\circ)}{1.65} = 0.549 \quad \text{or} \quad \theta_2 = 33.3^\circ$$

From geometry, $\phi_2 = A - \theta_2 = 60^\circ - 33.3^\circ = 26.7^\circ$. Snell's law applied to the second surface thus gives

$$\sin\phi_1 = \frac{n_2 \sin\phi_2}{n_1} = \frac{(1.65)(\sin 26.7^\circ)}{1.00} = 0.741 \quad \text{or} \quad \phi_1 = 47.8^\circ$$

The angle of deviation is

$$\theta_D = \theta_1 + \phi_1 - A = 65^\circ + 47.8^\circ - 60^\circ = 52.8^\circ$$

We now repeat the above steps with $n_2 = 1.68$ instead of 1.65. The results are

$$\theta_2 = 32.6^\circ \quad \phi_2 = 27.4^\circ \quad \phi_1 = 50.7^\circ \quad \theta_D = 55.7^\circ$$

Consequently, the difference in $\theta_D$ for blue and red light is $\Delta\theta_D = 55.7^\circ - 52.8^\circ = 2.9^\circ$.

## Chapter 26: Mirrors, Lenses, and Optical Instruments

**410.** The basic equation can be written as

$$\frac{1}{p}+\frac{1}{q}=\frac{1}{f} \quad (1)$$

where $p$ = distance of object, $q$ = distance of image, and $f$ = focal length, and distances are measured to the lens or mirror along the symmetry axis. This formula is valid only for thin lenses and mirrors whose dimensions are small compared with the radii of curvature involved, and for light rays coming in at shallow angles to the symmetry axis (paraxial rays).

The sign conventions for $p, q, f$ are given in Table A26.1. Light rays diverge from a point on a *real object* but converge to a point on a *virtual object.* Similarly, light rays converge toward a point on a *real image* but diverge from a point on a *virtual image.* Other symbols commonly used instead of $(p, q)$ are $(p, p')$, $(s, s')$, $(s_0, s_i)$, $(d_o, d_i)$, etc.

**TABLE A26.1**

| | Sign | |
|---|---|---|
| **Quantity** | **+** | **−** |
| $f$ | Converging lens, concave mirror | Diverging lens, convex mirror |
| $p$ | Real object | Virtual object |
| $q$ | Real image | Virtual image |

**411.** Real rays $a$, $b$, $c$ are drawn from a point $P_1$ on the real object $P_1P_2$. For convenience, $a$ is drawn parallel to the optic axis, $b$ through $F$, and $c$ through $C$. From the geometry of the mirror and the law of reflection, reflected ray $a'$ passes through $F$, $b'$ is parallel to the optic axis, and $c'$ returns through $C$ along the path of $c$. The intersection of the real rays $a'$, $b'$, $c'$ at $P_1'$ locates one point on the real image $P_1'P_2'$.

Rays $a$, $b$, $c$ may seem very special. However, any ray from $P_1$, after reflection, passes through $P_1'$, as may be shown by an application of the law of reflection. Note that the image is inverted.

Let $f = +20$ cm, $d_o = +45$ cm, $l_o = +5$ cm. Then

$$\frac{1}{45}+\frac{1}{d_i}=\frac{1}{20} \quad \text{or} \quad d_i = +36\text{ cm} \quad \text{(a real image)}$$

$$\frac{45}{36}=-\frac{5}{l_i} \quad \text{or} \quad l_i = -4\text{ cm} \quad \text{(an inverted, reduced image)}$$

**412.** See Figure A26.1. Constructing rays as in Figure A26.1, we find the image to be erect, virtual (behind the mirror), and magnified.

Letting $f = +20$ cm, $d_o = +15$ cm, $l_o = +5$ cm, we have from the lens equations $d_i = -60$ cm (virtual) and $l_i = +20$ cm (erect, magnified).

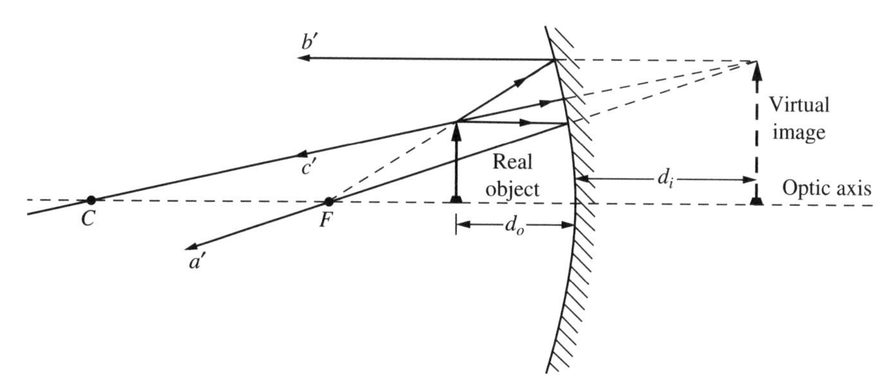


**Figure A26.1**

**413.** Sunlight consists of rays parallel to the principal axis. After reflection, the rays must all pass through the principal focus (Figure A26.2): $f = R/2 = 0.80/2 = 0.40$ m (focal length).

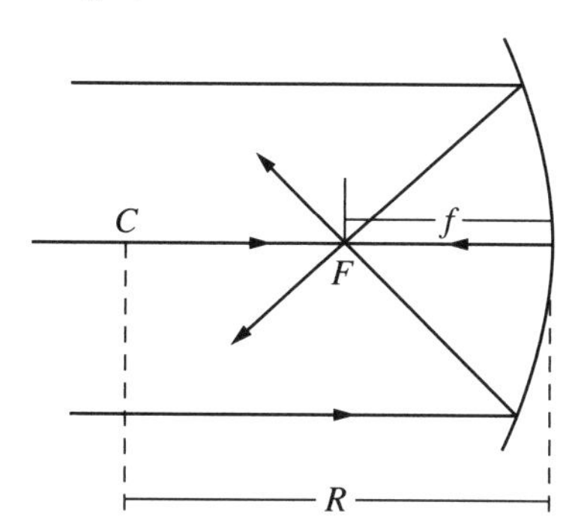


**Figure A26.2**

**414. (A)** $\frac{1}{p}+\frac{1}{q}=\frac{1}{f}$ $\quad \frac{1}{0.15}+\frac{1}{q}=\frac{1}{0.20}$ $\quad q=-0.60$ m

The image is virtual.

**(B)** $\frac{\text{height of image}}{\text{height of object}}=\left|\frac{q}{p}\right|$ $\quad \frac{\text{height of image}}{10 \text{ cm}}=\frac{0.6}{0.15}$ $\quad$ height of image = 40 cm

**415.** $\frac{1}{p}+\frac{1}{q}=\frac{1}{f}$ or $\frac{1}{50}+\frac{1}{q}=\frac{1}{20}$ and $q=\frac{100}{3}=33$ cm

$\frac{h_i}{h_o}=-\frac{q}{p}=\frac{-33.3}{50}=-0.666$ and $h_i=-6.7$ cm

Thus, image is real, 6.7 cm high and inverted.

**416.** $\dfrac{1}{p}+\dfrac{1}{q}=\dfrac{1}{f}$

If image and object are the same size, the magnification is 1. Then,

$$\left|\frac{q}{p}\right|=1 \qquad \text{and} \qquad q=p$$

Substituting $p$ for $q$ in the mirror formula and solving, $p = 2f$. The object must be placed $2f$ from the mirror. This point is the center of curvature. Note that $q = -p$ cannot satisfy the mirror formula.

**417.** Use the mirror formula with the proper signs.

$$\frac{1}{p}+\frac{1}{q}=\frac{1}{f} \qquad \frac{1}{0.5}+\frac{1}{-2.0}=\frac{1}{f} \qquad 2-\frac{1}{2}=\frac{1}{f}$$

$$\frac{3}{2}=\frac{1}{f} \qquad f=\frac{2}{3}\text{ ft} \qquad \frac{R}{2}=f\frac{2}{3}\text{ ft} \qquad R=1.3\text{ ft}$$

**418.** The focal length is $R/2$ = 18 cm. Then

$$\frac{1}{p}+\frac{1}{q}=\frac{1}{18} \qquad \text{and} \qquad \frac{1}{9}=\left|\frac{q}{p}\right| \qquad \text{with } q \text{ and } p \text{ positive (real object and image)}$$

Then $p = 9q$, so

$$\frac{1}{p}+\frac{9}{p}=\frac{1}{18} \qquad \text{or} \qquad \frac{10}{p}=\frac{1}{18} \qquad \text{and} \qquad p=180\text{ cm}$$

**419.** To cast the image on a wall, it must be real, so $q$ is positive, as is $p$, the object distance. Thus, the image is inverted and $q/p = 5$. Since $q > p$ we have, from the information given, $q - p = 12$ m, or $5p - p = 12$ m, and $p = 3$ m. The mirror is thus 3 m from the lamp and 15 m from the wall. Furthermore, $(1/p) + (1/q) = (1/f)$, or $\frac{1}{3}+\frac{1}{15}=1/f$, and $f = 2.5$ m. Thus, the mirror is concave of radius 5 m.

**420.** $f=\dfrac{r}{2}=\dfrac{0.2\text{ m}}{2}=0.1\text{ m}$ Then $\dfrac{1}{p}+\dfrac{1}{q}=\dfrac{1}{f}$ $\quad \dfrac{1}{0.5}+\dfrac{1}{q}=\dfrac{1}{0.1}$

$\dfrac{1}{q}=10-2=8$ $\quad q=0.125$ m from the mirror. For height we have

$$\frac{-q}{p}=\frac{h_i}{h_o} \qquad \frac{-0.125}{0.5}=\frac{h_i}{0.012} \qquad \text{and} \qquad h_i=-0.0030\text{ m}=-3.0\text{ mm}$$

The image is real and inverted.

**421.** See Figure A26.3. The image is real for $q$ positive, or $p > f$, and virtual for $q$ negative, or $p < f$.

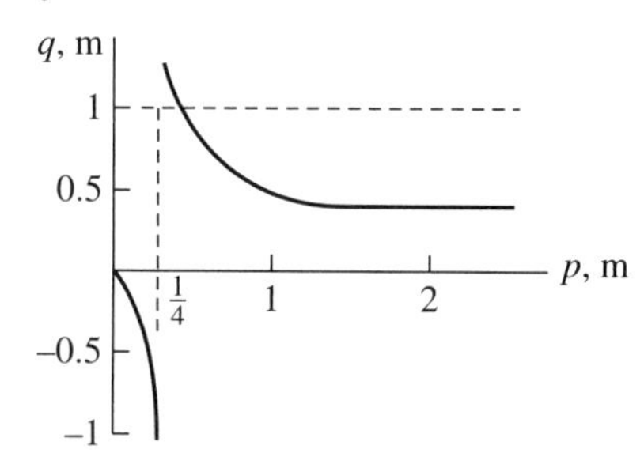


**Figure A26.3**

**422.** Since $p < f$, the image is virtual. Quantitatively,

$$\frac{1}{125}+\frac{1}{q}=\frac{1}{200} \qquad \frac{1}{q}=\frac{5}{1000}-\frac{8}{1000} \qquad \text{and}$$

$$q=\frac{-1000}{3}=-333 \text{ mm} \qquad \text{Then} \qquad \frac{-q}{p}=\frac{h}{8}=\frac{333}{125} \qquad h=21 \text{ mm}$$

The image is erect and virtual.

**423.** Applying the mirror equation, we have $\frac{1}{12}+1/q=\frac{1}{18}$, or $q=-36$ in. The image is virtual and upright with magnification $= |36/12| = 3.0$.

**424.** For magnification of the face, we are looking for a virtual upright image. $|q/p| = 2.5$, and thus $q = -2.5p = -1.0$ m. Next we find $f$ using

$$\frac{1}{0.4}+\frac{1}{-1.0}=\frac{1}{f} \qquad 2.5-1=\frac{1}{f} \qquad \text{and}$$

$$f=\frac{1}{1.5} \qquad \text{Finally} \qquad R=2f=2\left(\frac{1}{1.5}\right)=1.33 \text{ m}$$

**425.** We have

$$\frac{1}{9}+\frac{1}{q}=\frac{1}{12} \qquad \frac{1}{q}=\frac{3}{36}-\frac{4}{36}=-\frac{1}{36} \qquad \text{and} \qquad q=-36 \text{ mm}$$

The image is upright with

$$\text{magnification} = |36/9| = 4.0.$$

**426.** For the real image, $q = 3p$, and $1/p + 1/(3p) = 1/f = 1/0.15$, or $p = 0.20$ m.
For the virtual image, $q = -3p$ and $1/p + 1/(-3p) = 1/f = 1/0.15$, or $p = 0.10$ m.

In a facial magnifier, one wants a virtual image not only because it is upright but also because a magnified real image is behind one's face and one can't see it.

**427.** We find the image distance for each case. $1/375 + 1/q = 1/250$; $q = 750$ mm for $p = 375$ mm. Next, for $p$ 5 mm farther, $1/380 + 1/p = 1/250$; $q = 731$ mm for $p = 380$ mm. Thus, $\Delta q = -19$ mm, so the image moves in 19 mm.

**428. (A)** $f = R/2 = (-90)/2 = -45$ cm. Focal length $f$ must be negative for a convex mirror.

$$\frac{1}{p}+\frac{1}{q}=\frac{1}{f} \qquad \frac{1}{70}+\frac{1}{q}=-\frac{1}{45} \qquad q=-27 \text{ cm}$$

The image is 27.4 cm behind the mirror, and it is virtual.

**(B)** $-q/p = 27.4/70 = 0.39$ (magnification), and the image is upright.

**429.** We use the mirror equation $1/p + 1/q = 1/f$, which yields, recalling that $f$ is negative,

$$\frac{1}{24}+\frac{1}{q}=\frac{1}{-12} \quad \text{or} \quad \frac{1}{q}=-\frac{1}{12}-\frac{1}{24}=-\frac{3}{24} \quad \text{and} \quad q=-8 \text{ cm}$$

**430.** We have $|q/p| = \frac{1}{6}$. Since for a real object the image in a convex mirror is always virtual, we have $q = -p/6 = -\frac{12}{6} = -2$ cm. Then

$$\frac{1}{12}+\frac{1}{-2}=\frac{1}{f} \quad \text{or} \quad \frac{1}{12}-\frac{6}{12}=\frac{-5}{12}=\frac{1}{f} \quad \text{and} \quad f=-2.4 \text{ cm}$$

**431.** Recalling that object distance is measured to the reflecting surface and that $f = R/2$, we have

$$p=175-50=125 \text{ mm} \quad \text{and} \quad f=-25 \text{ mm} \quad \text{Then} \quad \frac{1}{125}+\frac{1}{q}=\frac{1}{-25}$$

$$\frac{1}{q}=-\frac{5}{125}-\frac{1}{125} \quad \text{and} \quad q=-\frac{125}{6}=-20.8 \text{ mm}$$

$$\text{linear magnification}=\left|\frac{q}{p}\right|=\frac{20.8}{125}=0.17$$

**432.** For a convex mirror, we have

$$f=\frac{r}{2}=\frac{-0.32}{2}=-0.16 \text{ m} \quad \text{Then} \quad \frac{1}{0.48}+\frac{1}{q}=\frac{1}{-0.16}$$

or

$$\frac{1}{q}=\frac{-3}{0.48}-\frac{1}{0.48}=-\frac{4}{0.48} \quad \text{and} \quad q=-0.12 \text{ m}$$

Magnification $= -q/p$, so $h/28 = 120/480$; $h = 7$ mm. The image is virtual and erect.

**433.** The sphere acts as a convex mirror with $f = R/2 = -\frac{8}{2} = -4$ in. Then, for the first part, with $p = 10$ in,

$$\frac{1}{10}+\frac{1}{q}=\frac{1}{-4} \qquad \frac{1}{q}=-\frac{5}{20}-\frac{2}{20} \qquad \text{and} \qquad q=-\frac{20}{7}=-2.9 \text{ in}$$

For the second part, we require $q = -16$ in. Then

$$\frac{1}{p}+\frac{1}{-16}=\frac{1}{-4} \qquad \text{or} \qquad \frac{1}{p}=\frac{1}{16}-\frac{4}{16}=\frac{-3}{16} \qquad \text{and} \qquad p=-5.3 \text{ in}$$

This cannot happen for a real object, so there is no location for her eye that will create the required image. This could also be seen by noting that for $p > 0$, $q$ must be negative and $|q| < |f|$. Thus, no real object has an image farther than $|f|$ behind the mirror.

**434.** In Figure A26.4, real rays $a$, $b$, $c$ from $P_1$ on real object $P_1P_2$ are, for convenience, drawn parallel to the optic axis, through $F$, and through the center of the lens. Refracted ray $a'$ passes through the back focal point $F'$, $b'$ is parallel to the optic axis, and $c'$ is the continuation of $c$. These rays, of course, follow the law of refraction. The intersection of real rays $a'$, $b'$, $c'$ locates point $P_1'$ on the real image $P_1'P_2'$. Other points on $P_1'P_2'$ can be located in the same way. The image is inverted.

Letting $f = +20$ cm, $d_o = +30$ cm, $l_o = +6$ cm, we find from the equation in Question 410 that $d_i = +60$ cm (real) and $l_i = -12$ cm (inverted). For the assumed values, the image is magnified.

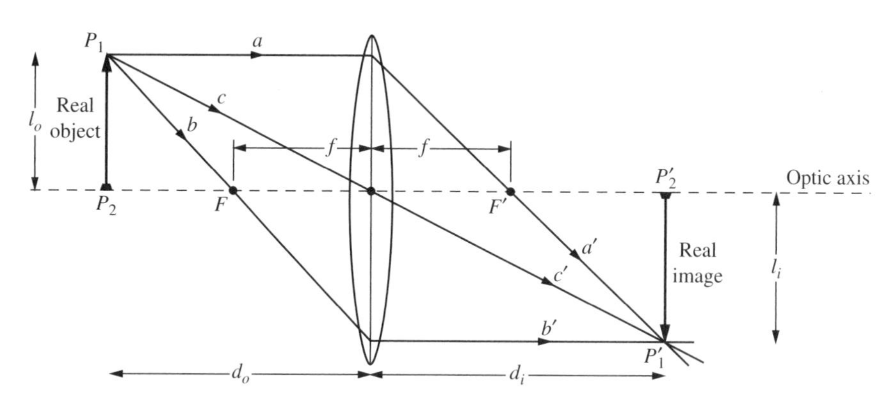


**Figure A26.4**

**435.** Figure A26.5 gives the ray construction; the image is virtual, erect, and magnified. Letting $f = +15$ cm, $d_o = +10$ cm, $l_o = 2$ cm in the equations Question 410, we find that $d_i = -30$ cm (virtual) and $l_i = +6$ cm (erect, magnified).

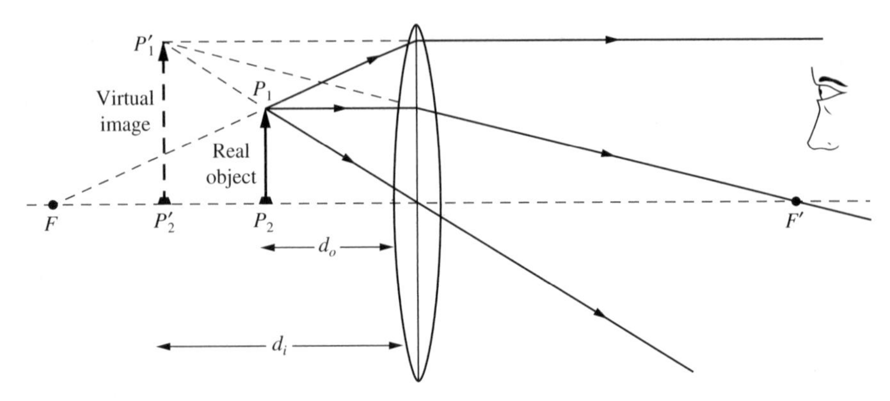


**Figure A26.5**

**436.** Figure A26.6 gives the ray construction, which is independent of the relationship between $d_o$ and $f$. The image is virtual, erect, and minified.

Checking with $f = -30$ cm, $d_o > 0$, $l_o = 10$ cm,

$$\frac{1}{d_o} + \frac{1}{d_i} = -\frac{1}{30} \quad \text{or} \quad d_i = -\frac{30d_o}{30 + d_o} \text{(virtual)} \quad \text{and}$$

$$\frac{30 + d_o}{30} = -\frac{10}{l_i} \quad \text{or} \quad l_i = \frac{10}{1 + (d_o/30)} \text{(erect, minified)}$$

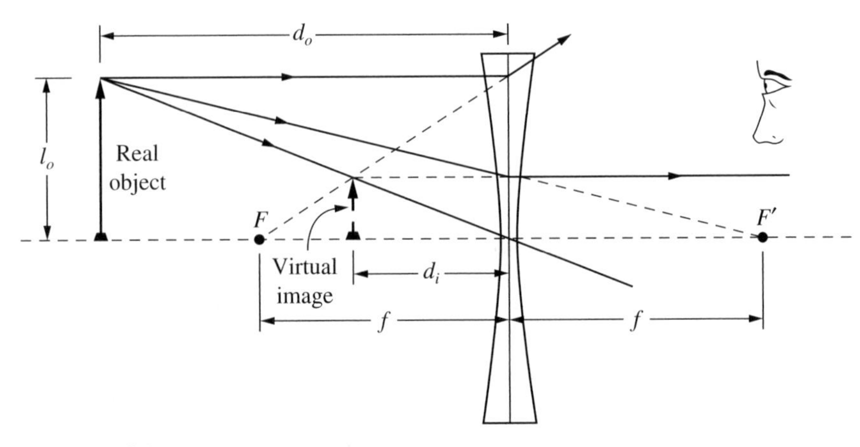


**Figure A26.6**

**437.** Let $p$, $p'$ be the object and image distances, respectively. Then we have $1/p + 1/p' = 1/f$, and $h_i/h_o = (-p')/p$. From the data, $f = 20$ cm for each case. Case 1: $\frac{1}{100} + 1/p' = \frac{1}{20}$; $p' = 25$ cm; image is real and inverted. Case 2: $\frac{1}{25} + 1/p' = \frac{1}{20}$; $p' = 100$ cm; image is real and inverted. Case 3: $\frac{1}{10} + 1/p' = \frac{1}{20}$; $p' = -20$ cm; image is virtual and erect.

**438.** Proceed as in Question 437. Case 1: $\frac{1}{30} + 1/p' = \frac{1}{10}$; $p' = 15$ cm; image is real and inverted. Case 2: $\frac{1}{5} + 1/p' = \frac{1}{10}$; $p' = -10$ cm; image is virtual and erect.

**439.** In this case, $f = -10$ cm. Images in both cases will be virtual and erect. Case 1: $\frac{1}{30} + 1/p' = -\frac{1}{10}$ so that $p' = -7.5$ cm. Case 2: $\frac{1}{5} + 1/p' = -\frac{1}{10}$, yielding $p' = -3.3$ cm. Image is virtual and erect.

**440.** We are given $3 = h_i/h_o = -q/p$, with $p$ positive (real object) and $q$ negative (virtual image). Then $q = -3p$.

$$\frac{1}{p}+\frac{1}{q}=\frac{1}{f} \quad \text{becomes} \quad \frac{1}{p}-\frac{1}{3p}=\frac{1}{f} \quad \text{or} \quad \frac{2}{3p}=\frac{1}{f} \quad \text{and} \quad p=\frac{2f}{3}$$

**441.** Since the image is one-third the size of the object and the image is always virtual for a diverging lens with real object,

$$\frac{q}{p}=-\frac{1}{3} \quad \text{and} \quad \frac{1}{p}+\frac{1}{q}=\frac{1}{f} \quad \text{yields} \quad \frac{1}{24}+\frac{1}{-8}=\frac{1}{f}$$

$$\frac{1}{24}-\frac{3}{24}=\frac{1}{f} \qquad -\frac{2}{24}=\frac{1}{f} \qquad f=-12 \text{ cm}$$

**442.** A projected image is real, so,

$$\frac{1}{p}+\frac{1}{q}=\frac{1}{f} \quad \text{becomes} \quad \frac{1}{10}+\frac{1}{90}=\frac{1}{f} \quad \text{or} \quad \frac{10}{90}=\frac{1}{f} \quad \text{and} \quad f=9 \text{ in}$$

**443.** The image must be virtual and $q$ is negative. Thus,

$$\frac{1}{p}+\frac{1}{q}=\frac{1}{f} \qquad \frac{1}{30}+\frac{1}{-90}=\frac{1}{f} \quad \text{or} \quad \frac{3}{90}-\frac{1}{90}=\frac{1}{f}$$

$$\frac{2}{90}=\frac{1}{f} \quad \text{and} \quad f=45 \text{ cm}$$

**444.** From $p + q = 10$ and $q = 4p$, we find $p = 2$ and $q = 8$. Then

$$\frac{1}{f}=\frac{1}{p}+\frac{1}{q}=\frac{1}{2}+\frac{1}{8}=\frac{5}{8} \quad \text{or} \quad f=\frac{8}{5}=+1.6 \text{ m}$$

**445.** From the lens equation,

$$\frac{1}{p}+\frac{1}{q}=\frac{1}{f} \qquad \frac{1}{25}+\frac{1}{q}=\frac{1}{20}$$

$$\frac{1}{q}=\frac{5}{100}-\frac{4}{100}=\frac{1}{100} \qquad q=100 \text{ cm}=1 \text{ m}$$

**446.** The linear magnification is to be $m = 100/5 = 20 = |q/p|$. Since object and image are real, $p$ and $q$ are positive, so $q = 20p$. Then $1/p + 1/q = 1/f$ becomes $20/p + 1/q = \frac{1}{15}$; $21/q = \frac{1}{15}$, and $q = 315$ cm.

## Chapter 27: Interference, Diffraction, and Polarization

**447.**

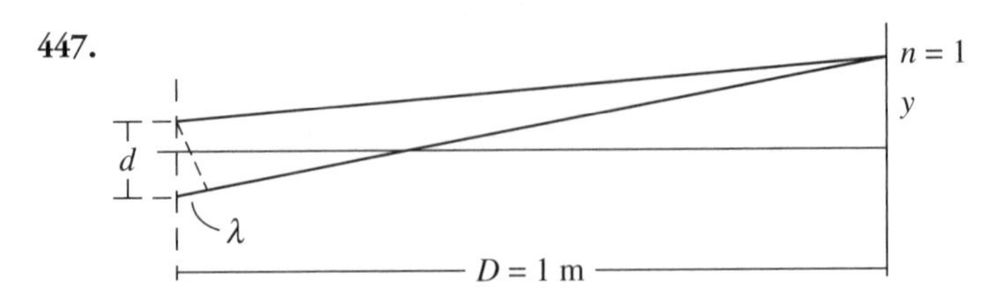


**Figure A27.1**

The situation is shown in Figure A27.1. For the first bright band, $n = 1$. The path difference is one wavelength.

$$n\lambda = \frac{dy}{D} \quad 5893\times10^{-10} = \frac{0.25\times10^{-3}\,y}{1} \quad y = \frac{5893\times10^{-10}}{0.25\times10^{-3}} = 2.36\times10^{-3}\text{ m} = 2.36\text{ mm}$$

Separation of adjacent bright fringes.

**448.** With $y = x_n$ the position of the $n$th maximum on the screen, measured from the central maximum. The distance between adjacent fringes is

$$\Delta x = x_{n+1} - x_n = \frac{D\lambda}{d}(n+1) - \frac{D\lambda}{d}n = \frac{D\lambda}{d}$$

so
$$d = \frac{D\lambda}{\Delta x} = \frac{(0.8\text{ m})(589\times10^{-9}\text{ m})}{0.35\times10^{-2}\text{ m}} = 1.35\times10^{-4}\text{ m} = 0.135\text{ mm}$$

**449.** From $\lambda = (d/D)\Delta x$, where $d$ and $D$ are fixed,

$$\lambda' = \frac{\Delta x'}{\Delta x}\lambda = \frac{7.6}{8.3}630 = 577\text{ nm}$$

**450.** Path difference $= 4\lambda = 4(6000\text{ Å}) = 24000\text{ Å}$ or $2.4\ \mu\text{m}$

**451.** Apply the condition for constructive interference twice: $(4\lambda_1)/d = x/D$ and $(6\lambda_2)/d = x/d$, so that $4\lambda_1 = 6\lambda_2$. If $\lambda_2 = 430$ nm, then $\lambda_1 = 645$ nm; if $\lambda_1 = 430$ nm, then $\lambda_2 = 287$ nm.

**452.** Inside the soap film, the wavelength of light $\lambda_f$ is shorter than it is in air by a factor of the index of refraction of the soap film, since the velocity of light is less inside the film.

$$\lambda_f = \frac{\lambda_a}{n} = \frac{500}{1.333} = 375\text{ nm}$$

The film thickness is $\frac{1}{4}\lambda_f$ for constructive interference, including the phase change on reflection at the top surface: $\frac{1}{4}\lambda_f = \frac{375}{4} = 94$ nm.

**453.** The pattern is caused by interference between a beam reflected from the upper surface of the air wedge and a beam reflected from the lower surface of the wedge. The two reflections are of different nature in that reflection at the upper surface takes place at the boundary of a medium (air) of lower refractive index, while reflection at the lower surface occurs at the boundary of a medium (glass) of higher refractive index. In such cases, the reflection involves a phase displacement of 180° between the two reflected beams. This explains the presence of a dark fringe at the left-hand edge.

As we move from a dark fringe to the next dark fringe, the beam that traverses the wedge must be held back by a path-length difference of $\lambda$. Because the beam travels twice through the wedge (down and back up), the wedge thickness changes by only $\frac{1}{2}\lambda$ as we move from fringe to fringe.

$$\text{spacer thickness} = 4\left(\frac{1}{2}\lambda\right) = 2(589\text{ nm}) = 1180\text{ nm}$$

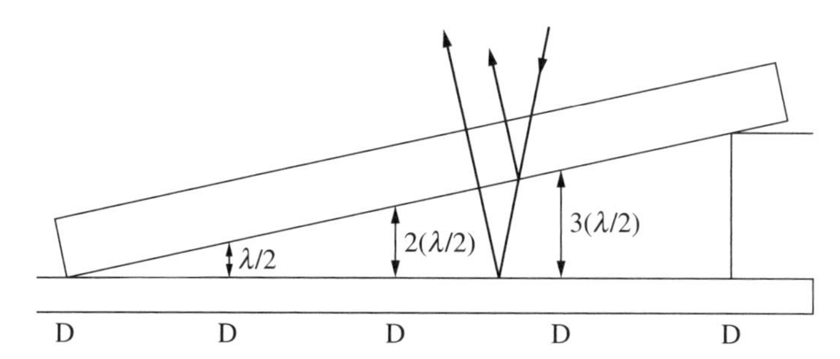


**Figure A27.2**

**454.** The $n$th-order line obeys $n\lambda = d\sin\theta$. For our case, $n = 1$ and $d = \frac{1}{6000}$ cm, so $\lambda = \frac{1}{6000}\sin 30° = 833\times10^{-7}\text{ cm} = 833\text{ nm}$.

**455.** Use the grating formula and solve for $\theta$ for both $n = 1$ and $n = 2$, with distances in meters:

$$n\lambda = d\sin\theta \qquad 5461\times10^{-10} = \frac{1}{(39.37)(15000)}\sin\theta_1 \qquad \text{or}$$

$$0.3225 = \sin\theta_1 \qquad \text{and} \qquad \theta_1 = 18.8° \text{ first order}$$

Next, for $n = 2$,

$$2(5461\times10^{-10}) = \frac{1}{(39.37)(15000)}\sin\theta_2 \qquad \text{or} \qquad 0.6449 = \sin\theta_2 \qquad \text{and}$$

$$\theta_2 = 40.2° \text{ second order}$$

Finally, $\theta_2 - \theta_1 = 40.2° - 18.8° = 21.4°$ separation.

**456.** According to the grating equation, the long-wavelength limit of the $m$th-order spectrum and the short-wavelength limit of the $(m + 1)$th-order spectrum will just coincide if

$$d\sin\theta = m(600\times10^{-9}) = (m+1)(450\times10^{-9})$$

whence $m = 3$. Then, since $\theta = 30°$,

$$\frac{1}{d} = \frac{\sin 30°}{3(600\times10^{-9})} = 280{,}000 \text{ lines per meter}$$

**457.** We apply Malus's law.

$$I = 0.1I_0 = I_0 \cos^2\theta \quad \text{so} \quad \cos^2\theta = 0.1 \quad \text{Then} \quad \cos\theta = \sqrt{0.1} = 0.316 \quad \text{and} \quad \theta = 72°$$

**458.** The ratio of the transmitted intensity $I'$ to the incident intensity $I$ is, by Malus's law,

$$\frac{I'}{I} = \cos^2\theta = \cos^2 65° = 0.18$$

**459.** The unpolarized light behaves like two equal-intensity polarized waves polarized at right angles to each other. We can assume one is along the incident transmission axis, and hence the transmitted light has intensity $\frac{1}{2}I'$. This light now passes through the second filter, and by Malus's law $I = \left(\frac{1}{2}I'\right)\cos^2\theta$.

**460. (A)** From Malus's law for the second Polaroid, $I_1/I_2 = \cos^2\phi = \cos^2 30° = 0.866^2 = 0.75$. Three-fourths of the light striking the second Polaroid is transmitted. The first Polaroid removes half the unpolarized light.

Therefore, $\frac{1}{2} \times \frac{3}{4} = \frac{3}{8}$ of the original light transmitted.

**(B)** Now the final intensity is 20 percent of the intensity of the light entering the second Polaroid, so $I_1/I_2 = 0.2 = \cos^2\phi$, and $\phi = 63°$.

**461.** Since the angle between the axis of the first analyzer and the initial amplitude $\mathbf{A}_0$ is 45°, the intensity after passing through the first analyzer is $I' = I_0 \cos^2 45° = 0.5I_0$. The transmitted amplitude $\mathbf{A}'$ is oriented at 45° with respect to the axis of the second analyzer, so the final intensity is $I = I' \cos^2 45° = 0.5I' = 0.25I_0$, and the final amplitude $\mathbf{A}$ is oriented at 90° with respect to the initial amplitude $\mathbf{A}_0$.

Note that if only the second analyzer were in place, *no* light would get through, since $\mathbf{A}_0$ is perpendicular to its transmission axis. This problem thus demonstrates the vector nature of the light disturbance.

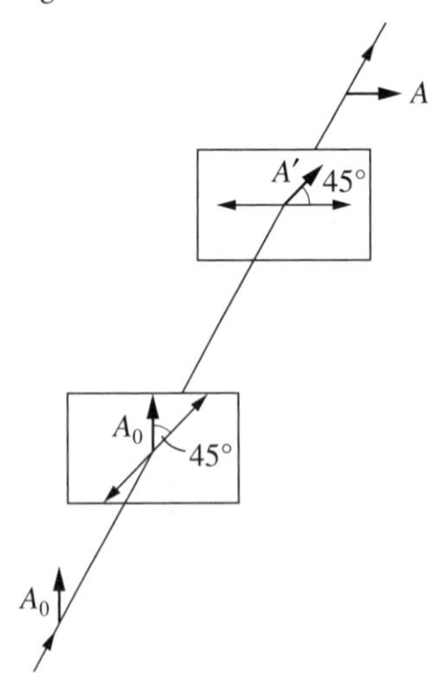


**Figure A27.3**

## Chapter 28: Special Relativity

**462.** Frame of spaceship:

$$l_0 = 100 \text{ m}$$
$$t_0 = ?$$
$$v_0 = \frac{l_0}{t_0}$$

Frame of earth:

$$l = ?$$
$$t = 4\ \mu\text{s}$$
$$v = \frac{l}{t}$$

Length contraction:

$$l = l_0\sqrt{1-\frac{v^2}{c^2}}$$

so $v = \dfrac{l_0\sqrt{1-\frac{v^2}{c^2}}}{t}$ solving for $v$,

$v = \dfrac{l_0}{t}\left(\dfrac{1}{\sqrt{1+\frac{l_0^2}{t^2c^2}}}\right)$ plugging in values, the parenthetical term is 0.961

$$v = 2.5\times10^8\ \frac{\text{m}}{\text{s}}(.961) = 2.4\times10^8\ \frac{\text{m}}{\text{s}}$$

**463.** Some sort of timing mechanism within the particle determines how long it "lives." This internal clock, which gives the proper lifetime, must obey the time dilation relation. We have

$$t_p = t_0\sqrt{1-(v/c)^2} \qquad \text{or} \qquad 0.75\times10^{-8} = (2\times10^{-8})\sqrt{1-(v/c)^2}$$

Note that $t_p$ is the time that the moving clock ticks out during the $2 \times 10^{-8}$ s ticked out by the laboratory clock. Squaring each side of the equation and solving for $v$ gives $v = 0.927c =$ $2.78 \times 10^8$ m/s.

**464.** The time for doubling, $t_0$, as seen on earth is given by $t_0 = t_p/\sqrt{1-(v^2/c^2)}$, where $t_p = 20$ days is the proper doubling time. Then $t_0 = 20/\sqrt{1-0.9950^2} = 200$ days. Thus, in 1000 earth days 5 doublings take place. Starting from 2 bacteria there are $2^6$, or 64 bacteria.

**465.** The stick behaves normally; it does not appear to change its length, because it has no translational motion relative to the observer in the spaceship. However, an observer on earth would measure the stick to be $(1\text{ m})\sqrt{1-(v/c)^2}$ long when parallel to the ship's motion, and 1 m long when perpendicular to the ship's motion.

**466.** Let $L$ be the length of the line as seen in the first spaceship. The horizontal and vertical extents of the line are each $L/\sqrt{2}$. In the second ship, the horizontal extent is shortened to $(L/\sqrt{2})\sqrt{1-(v^2/c^2)}$ while the vertical extent is unchanged. Then

$$\tan\theta = \frac{\text{horizontal}}{\text{vertical}} = \sqrt{1-\frac{V^2}{c^2}} \qquad \text{or} \qquad \tan\theta = \sqrt{1-0.95^2} = 0.31 \qquad \text{and} \qquad \theta = 17.3°$$

**467.** Let observers $O$, $O'$ be associated with the earth and rocket $B$, respectively. Then, let $u_x$ be the velocity of rocket $A$ relative to the earth,

$$u'_x = \frac{u_x - v}{1-(v/c^2)u_x} = \frac{0.8c-(-0.6c)}{1-[(-0.6c)(0.8c)]/c^2} = 0.946c$$

The problem can also be solved with other associations. For example, let observers $O$ and $O'$ be associated with rocket $A$ and rocket $B$, respectively. Let $u_x$, $u'_x$ represent velocity of earth. Then

$$u'_x = \frac{u_x - v}{1-(v/c^2)u_x} \qquad \text{or} \qquad 0.6c = \frac{-0.8c - v}{1-(v/c^2)(-0.8c)}$$

Solving, $v = -0.946c$, which agrees with the above answer. (The minus sign appears because $v$ is the velocity of $O'$ with respect to $O$, which, with the present association, is the velocity of rocket $B$ with respect to rocket $A$.)

**468.** The galilean transformation is $u'_{xG} = u_x - v$ and the Lorentz transformation is

$$u'_{xR} = \frac{u_x - v}{1-(v/c^2)u_x} = \frac{u'_{xG}}{1-(v/c^2)u_x}$$

Rearranging,

$$\frac{u'_{xR} - u'_{xG}}{u'_{xR}} = \frac{vu_x}{c^2}$$

Thus, if the product $vu_x$ exceeds $0.02c^2$, the error in using the galilean transformation instead of the Lorentz transformation will exceed 2 percent.

**469.** *Method 1* $c$ (by the second postulate of special relativity)

*Method 2* According to the velocity addition formula, the observed speed will be (since $u = c$ in this case)

$$\frac{v+u}{1+(uv/c^2)} = \frac{v+c}{1+(v/c)} = \frac{(v+c)c}{c+v} = c$$

**470.** $m = m_0[1-(v^2/c^2)]^{-1/2} \approx m_0[1+\frac{1}{2}(v^2/c^2)]$, or $(m-m_0) = \frac{1}{2}m_0(v^2/c^2) = \frac{1}{2}(2000)\times [15/(3.0\times10^8)]^2 = 2.5\times10^{-12}$ kg

**471.** The kinetic energy is $K = mc^2 - m_0c^2$, or $1.6\times10^{-14}$ J $= (m-m_0)(3\times10^8 \text{ m/s})^2$, and $m - m_0 = 1.78\times10^{-31}$ kg. We have $m_0 = 9.11\times10^{-31}$ kg; so $m = (9.11+1.78)\times10^{-31}$ kg $= 1.089\times10^{-30}$ kg. To get velocity, we note that $m = m_0/\sqrt{1-(v^2/c^2)}$, or $1-(v^2/c^2) = (m_0/m)^2 = 0.700$. Then $v/c = 0.548$ and $v = 1.64\times10^8$ m/s.

**472.** Use the mass-increase formula.

$$m = m_0\Big/\sqrt{1-\frac{v^2}{c^2}} \qquad 2m_0 = m_0\Big/\sqrt{1-\frac{v^2}{c^2}} \qquad 4\left(1-\frac{v^2}{c^2}\right)=1; \qquad \frac{v^2}{c^2}=\frac{3}{4} \qquad v = 0.866c$$

$$\begin{aligned}\Delta W &= (m-m_0)c^2 = (2m_0 - m_0)c^2 = (1.67\times10^{-27}\text{ kg})(3\times10^8\text{ m/s})^2\\ &= 1.50\times10^{-10}\text{ J} = 938\text{ MeV}\end{aligned}$$

**473.** Use the mass-increase formula.

$$m = \frac{m_0}{\sqrt{1-(v^2/c^2)}} = \frac{9.1\times10^{-31}\text{kg}}{\sqrt{1-0.95^2}} = 29.1\times10^{-31}\text{kg}$$

$$\Delta W = (m-m_0)c^2 = [(29.1-9.1)(10^{-31})](9\times10^{16}) = 1.8\times10^{-13}\text{ J} = 1.1\text{ MeV}$$

**474.** We use $\Delta E = (\Delta m)c^2$, with $\Delta E = mgh$. Therefore,

$$\Delta m = \frac{\Delta E}{c^2} = \frac{mgh}{c^2} = \frac{(2\text{ kg})(9.8\text{ m/s}^2)(0.3\text{ m})}{(3\times10^8\text{ m/s})^2} = 6.5\times10^{-17}\text{ kg}$$

**475.** $\Delta E = mC\,\Delta T = (100)(0.389)(100) = 3890$ kJ $\qquad \Delta m = \frac{\Delta E}{c^2} = 4.33\times10^{-11}$ kg

**476.** We make use of $\Delta E = (\Delta m)c^2$.

$$\text{energy gained} = (\text{mass lost})c^2 = (10^{-3}\text{kg})(3\times10^8\text{ m/s})^2 = 9\times10^{13}\text{ J}$$

$$\text{value of energy} = (9\times10^{13}\text{ J})\left(\frac{1\text{ kW}\cdot\text{h}}{3.6\times10^6\text{ J}}\right)\left(\frac{\$0.01}{\text{kW}\cdot\text{h}}\right) = \$250{,}000$$

## Chapter 29: Particles of Light and Waves of Matter

**477.** The wavelength $\lambda = c/f = (3.00\times10^8)/(9.00\times10^5) = 333$ m. The wave crests past a point corresponding to the number of wavelengths per second, the frequency $9.00\times10^5$ Hz (regardless of distance from the radio station).

**478.** $E = hf = (hc)/\lambda = [(6.63\times10^{-34}\ \text{J}\cdot\text{s})(3\times10^{8}\ \text{m/s})]/(526\times10^{-9}\ \text{m}) = 3.78\times10^{-17}\ \text{J}$. Since 1 eV = $1.6\times10^{-19}$ J, $E = 2.36$ eV.

Calculations like this are facilitated by using suitable mixed units for $hc$:

$$hc = (6.63\times10^{-34}\ \text{J}\cdot\text{s})(3\times10^{8}\ \text{m/s})(1.6\times10^{-19}\ \text{J/eV}) = 1240\ \text{eV}\cdot\text{nm}$$
$$= 1.240\ \text{MeV}\cdot\text{pm} = \cdots \qquad (1)$$

Thus, $E = (1240\ \text{eV}\cdot\text{nm})/(526\ \text{nm}) = 2.36$ eV.

**479.** Using the expression $E = hf$, we have $\mathscr{E} = (6.6\times10^{-34})(60) = 39.6\times10^{-33}$ J as the energy of a 60-Hz photon. Light extends from $3.8\times10^{14}$ Hz to $7.7\times10^{14}$ Hz, or from $25.1\times10^{-20}$ J to $50.8\times10^{-20}$ J. Thus, 60 Hz is less energetic than is light by a factor of about $10^{13}$.

**480.** The power multiplied by the given time interval is the emitted energy, that is, (100 W) × (1 s) = 100 J. Denoting the photon flux by $N$, we have

$$N = \frac{100}{hf} = \frac{100\lambda}{hc} = \frac{100(500\times10^{-9})}{(6.6\times10^{-34})(3\times10^{8})} = 25\times10^{19}\ \text{photons/s}.$$

**481.** The energy of a photon of the light is

$$E = \frac{hc}{\lambda} = \frac{(6.63\times10^{-34})(3\times10^{8})}{600\times10^{-9}} = 3.3\times10^{-19}\ \text{J}$$

The lamp uses 200 W of power. The number of photons emitted per second is therefore

$$n = \frac{200}{3.3\times10^{-19}} = 6.1\times10^{20}\ \text{photons/s}$$

Since the radiation is spherically symmetrical, the number of photons entering the sensor per second is $n$ multiplied by the ratio of the aperture area to the area of a sphere of radius 10 m:

$$(6.1\times10^{20})\frac{\pi(0.010^{2})}{4\pi(10^{2})} = 1.53\times10^{14}\ \text{photons/s}$$

and the number of photons that enter the sensor in 0.1 s is $(0.1)(1.53\times10^{14}) = 1.53\times10^{13}$.

**482.** The requirement is that $(mv)_{\text{electron}} = (h/\lambda)_{\text{photon}}$. From this,

$$\lambda = \frac{h}{mv} = \frac{6.63\times10^{-34}\ \text{J}\cdot\text{s}}{(9.1\times10^{-31}\ \text{kg})(2\times10^{5}\ \text{m/s})} = 3.64\ \text{nm}$$

This wavelength is in the x-ray region.

**483.** We know that $F = \Delta p/\Delta t$ and, in this case, $\Delta p$ is *twice* the incident momentum. Thus, if $N$ is the number of incoming photons per second, $F = N[(2h)/\lambda]$.

$$N = \frac{F\lambda}{2h} = \frac{663\times10^{-9}(1)}{2(6.63\times10^{-34})} = 5\times10^{26}\ \text{photons/s}$$

**484.** energy of photon $= \dfrac{hc}{\lambda} = \dfrac{1200 \text{ eV} \cdot \text{nm}}{200 \text{ nm}} = 6.20 \text{ eV}$

Then, from the photoelectric equation, the energy of the fastest-emitted electron is 6.20 eV – 5.00 eV = 1.20 eV. Hence, a retarding potential of 1.20 V is required.

**485.** At threshold, the photon energy just equals the energy required to tear the electron loose from the metal, namely, the work function $W_{\text{min}}$.

$$W_{\text{min}} = \frac{hc}{\lambda} \qquad \text{or} \qquad \lambda = \frac{1240 \text{ eV} \cdot \text{nm}}{2.3 \text{ eV}} = 540 \text{ nm}$$

**486.** As in Question 485,

$$\text{threshold } \lambda = \frac{hc}{W_{\text{min}}} = \frac{1240 \text{ eV} \cdot \text{nm}}{4.4 \text{ eV}} = 282 \text{ nm}$$

Hence, visible light (400 to 700 nm) cannot eject photoelectrons from copper.

**487.** **(A)** $E = \dfrac{hc}{\lambda} = \dfrac{1240 \text{ eV} \cdot \text{nm}}{600 \text{ nm}} = 2.07 \text{ eV}$

**(B)** $K_{\text{max}} = E - W_{\text{min}} = 2.07 - 2 = 0.07 \text{ eV}$

**(C)** $eV_s = K_{\text{max}} = 0.07 \text{ eV}$ or $V_s = 0.07 \text{ V}$

**488.** $W_{\text{min}} = hf - K_{\text{max}} = 4.0 - 1.1 = 2.9 \text{ eV}$

**489.** The first-order maximum is at $\lambda = d \sin \theta = d \sin 25°$. From de Broglie's relation, $h/(mv) = \lambda = (6.63 \times 10^{-34})/(9.1 \times 10^{-31} \times 400)$, or $\lambda = 1.82\ \mu\text{m}$. Then $d = \lambda/(\sin 25°) = 4.3\ \mu\text{m}$. If the maximum were $n$th order, then $n\lambda = d \sin \theta$ and $d = n(4.3\ \mu\text{m})$.

## Chapter 30: Modern Physics: Atoms, Nuclei, and Solid-State Electronics

**490.** $\Delta E = hf = \dfrac{hc}{\lambda} = \dfrac{1240 \text{ eV} \cdot \text{nm}}{589.6 \text{ nm}} = 2.1 \text{ eV}$

**491.** Using $E = (hc)/\lambda$ for the x-ray, we find that the x-ray energy = 1240/1.37 = 905 eV. The binding energy for the electron in question is 905 – 83 = 822 eV; the electron energy level is –822 eV.

**492.** $A - Z = 23 - 11 = 12$.

**493.** One atom of $^{12}\text{C}$ consists of six protons, six electrons, and six neutrons. The mass of the uncombined protons and electrons is the same as that of six $^{1}\text{H}$ atoms (if we ignore the very small binding energy of each proton-electron pair). The component particles may thus be considered as six $^1$H atoms and six neutrons. A mass balance may be computed as follows.

mass of six $^1$H atoms $= 6 \times 1.0078 = 6.0468$ u

mass of six neutrons $= 6 \times 1.0087 = 6.0522$ u

total mass of component particles $= 12.0990$ u

mass of $^{12}$C $= 12.0000$ u

loss in mass on forming $^{12}$C $= 0.0990$ u

binding energy $= (931 \times 0.0990)$ MeV $= 92$ MeV

**494.** Charge is conserved, so $Z = 92 - 57 = 35$.

**495. (A)** Each fission yields 200 MeV = $(200 \times 10^6)(1.6 \times 10^{-19})$ J of energy. Only 20 percent of this is utilized efficiently and so

$$\text{energy generated per fission} = (200 \times 10^6)(1.6 \times 10^{-19})(0.20) = 6.4 \times 10^{-12}\ \text{J}$$

Because the reactor's output is $700 \times 10^6$ J/s, the number of fissions required per second is

$$\text{fissions/s} = \frac{7 \times 10^8\ \text{J/s}}{6.4 \times 10^{-12}\ \text{J}} = 1.1 \times 10^{20}\ \text{s}^{-1} \qquad \text{and}$$

$$\text{fissions/day} = (86\,400\ \text{s/d})(1.1 \times 10^{20}\ \text{s}^{-1}) = 9.5 \times 10^{24}\ \text{per day}$$

**(B)** There are $6.02 \times 10^{26}$ atoms in 235 kg of uranium-235. Therefore, the mass of uranium-235 consumed in one day is

$$\text{mass} = \left(\frac{9.5 \times 10^{24}}{6.02 \times 10^{26}}\right)(235\ \text{kg}) = 3.7\ \text{kg}$$

**496.** We have

$$_0n^1 + {}_5\text{B}^{10} \rightarrow {}_3\text{Li}^7 + {}_2\text{He}^4$$

$$1.00866 + 10.01294 \rightarrow 7.01600 + 4.00260 \qquad 11.02160 \rightarrow 11.01860$$

$$11.02160 - 11.01860 = 0.00300\ \text{u converted to energy} \qquad 0.00300 \times 931 = 2.79\ \text{MeV}$$

**497. (A)** ${}^{222}_{86}\text{Rn} \rightarrow {}^{218}_{84}\text{Po} + {}^{4}_{2}\text{He}$

**(B)** ${}^{2}_{1}\text{H} + {}^{2}_{1}\text{H} \rightarrow {}^{3}_{1}\text{H} + {}^{1}_{1}\text{H}$

**(C)** ${}^{239}_{93}\text{Np} \rightarrow {}^{239}_{94}\text{Pu} + {}^{0}_{-1}\beta + \bar{v}$ (antineutrino)

**(D)** ${}^{22}_{11}\text{Na} \rightarrow {}^{22}_{10}\text{Ne} + {}^{0}_{+1}\beta + v$ (neutrino)

**498. (A)** The sum of the subscripts on the left is $7 + 2 = 9$. The subscript of the first product on the right is 8. Hence, the second product on the right must have a subscript (net charge) of 1. The sum of the superscripts on the left is $14 + 4 = 18$. The superscript of the first product is 17. Hence, the second product on the right must have a superscript (mass number) of 1. The particle with a nuclear charge 1 and a mass number 1 is the proton, ${}^{1}_{1}\text{H}$.

**(B)** The nuclear charge of the second product particle (its subscript) is $(4+2)-6=0$. The mass number of the particle (its subscript) is $(9 + 4) - 12 = 1$. Hence, the particle must be the neutron, ${}_{0}^{1}n$.

**499. (A)** The nuclear charge of the second particle is $15 - 14 = +1$. The mass is $30 - 30 = 0$. Hence, the particle must be a positron, ${}_{+1}^{0}e$.

**(B)** The nuclear charge of the second particle is $1 - 2 = -1$. Its mass number is $3 - 3 = 0$. Hence, the particle must be a beta particle (an electron), ${}_{-1}^{0}e$.

**500.** The activity is proportional to the number of undecayed atoms ($\Delta N/\Delta t = \lambda N$).

**(A)** In each half-life, half of the remaining sample decays. Because $\frac{1}{2} \times \frac{1}{2} \times \frac{1}{2} = \frac{1}{8}$, three half-lives, or 15.8 years, are required for the sample to decay to one-eighth its original strength.

**(B)** Take $\log_2(3) = 1.6$ to find that it will take about 1.6 half-lives, or 8.4 years.